Adaptive Beaming and Imaging in the Turbulent Atmosphere

Adaptive Beaming and Imaging in the Turbulent Atmosphere

Vladimir P. Lukin
Boris V. Fortes

Translated by A. B. Malikova

An update and translation of the Russian-language publication,
Adaptivnoe Formirovanie Puchkov i Izobrazhenii v Atmosfere
(Siberian Branch of the Russian Academy of Sciences, Novosibirsk, 1999).

SPIE PRESS
A Publication of SPIE—The International Society for Optical Engineering
Bellingham, Washington USA

Library of Congress Cataloging-in-Publication Data

Lukin, V. P. (Vladimir Petrovich)
[Adaptivnoe formirovanie puchkov i izobrazhenii v atmosfere. English]
Adaptive beaming and imaging in the turbulent atmosphere / V.P. Lukin and B.V. Fortes ; translated by A.B. Malikova.
p. cm. – (SPIE Press monograph ; PM109)
Includes bibliographical references and index.
ISBN 0-8194-4337-9 (softcover)
1. Laser beams—Atmospheric effects. 2. Atmosphere—Laser observations. 3. Atmospheric turbulence. 4. Optics, Adaptive. I. Fortes, B. V. (Boris V.) II. Title. III. Series.

QC976.L36 L8513 2002
621.36'9—dc21 2002070761

Published by

SPIE—The International Society for Optical Engineering
P.O. Box 10
Bellingham, Washington 98227-0010 USA
Phone: (1) 360.676.3290
Fax: (1) 360.647.1445
Email: spie@spie.org
Web: www.spie.org

Printed in the United States of America.

CONTENTS

PREFACE TO THE ENGLISH EDITION

This monograph is presented by two authors working at the Institute of Atmospheric Optics, Siberia, Russian Academy of Sciences (Tomsk, Russia). It is an overview of our results reported in recent years (up to 1999).

Specialists know that now it is practically impossible to write a book that thoroughly reviews the state of the art in adaptive optics because of the rapid advances in this field. Therefore, in this book we omit such a review and only give the necessary references to original papers (as of 1998, the year this book was written in Russian). We apologize to those authors whose papers were undeservedly ignored and not cited.

This book does not pretend to be a generalization of the most recent results. It is more like a compendium of results and ideas the authors followed when developing this particular area of modern optics. These days, the world community has developed approaches and concepts different from those presented in this book, and they have the right to their own existence and development as well.

With due respect for our readers,

Vladimir P. Lukin
Boris V. Fortes
June 2002

INTRODUCTION

The extensive use of optical technologies for solving problems of information transfer, narrow-directional electromagnetic energy transport, and image formation in an outdoor atmosphere calls for the development of adaptive correction methods and devices of that are an effective means of controlling the decrease in the efficiency of atmospheric optical systems caused by inhomogeneities in large-scale refractive indexes. These inhomogeneities are due to the turbulent mixing of atmospheric air masses and molecular and aerosol absorption in the channel of optical radiation propagation.

Adaptive optical systems (AOS) that operate in real time allow one to

- improve laser radiation focusing on a target, and hence increase the radiation intensity within the focal spot;
- decrease the image blooming of astronomical and other objects in telescopes, increase image sharpness, and decrease the probability of object recognition errors; and
- decrease the noise level and increase the data rate in optical communication systems.

Annual international conferences on adaptive optics held under the auspices of SPIE (The International Society for Optical Engineering), OSA (Optical Society of America), and adaptive optics sessions included in the programs of other conferences on atmospheric optics testify to the urgency of this problem. In 1994, a special issue of the *Journal of the Optical Society of America* was devoted to problems of adaptive correction of atmospheric distortions. Special annual issues of *Atmospheric and Oceanic Optics* are published by the Institute of Atmospheric Optics. Recently, AOS has been introduced in astronomical telescopes in many countries, including Russia, where the original Russian project of the AST-10 10-m adaptive telescope is being developed.

Wide practical application of AOS has revealed a number of problems that call for the development of a theory of optical wave propagation under adaptive control conditions. A search for answers to these problems necessitates the development of detailed and adequate mathematical AOS models and the application of research methods such as numerical experiments that solve a system of differential equations describing optical wave propagation in the atmosphere.

This monograph is primarily concerned with the original results of our investigations carried out using numerical experiments (models). The sole exceptions are sections devoted to adaptive image formation. Numerical experiments allow the maximum number of parameters to be considered to correctly model AOS and to investigate practically any significant radiative characteristic—the effective size of the light spot, the peak radiant intensity, the radiation power incident on the receiving aperture, the statistical characteristics of the radiant intensity and phase—in the context of a universal approach. A numerical experiment with applications to AOS allows one to predict the efficiency of various system configurations. Much time and considerable expense would be required to perform field experiments.

Work on numerical modeling of atmospheric distortions of beams and images and also on the possibility of their adaptive correction goes back to the early 1970s. It was started nearly simultaneously in several large U.S. laboratories (including Lincoln Laboratory at Massachusetts Institute of Technology and the Lawrence Livermore Laboratory at the University of California). Here it is pertinent to mention, in particular, the first reports on the results of numerical modeling of thermal self-action obtained by Gebhardt and Smith, Bradley and Herrmann, and Ulrich et al. The first work devoted to phase compensation for thermal blooming was published in 1974, and the first work devoted to a numerical investigation of the adaptive correction for turbulent image distortions was published in the same year. In 1976, Fleck, Morris, and Feit described in detail a procedure for solving the nonstationary problem of thermal self-action in a turbulent atmosphere. The first special issue of the *Journal of the Optical Society of America*, which summarized the results of theoretical and experimental investigations into adaptive optics in the United States, was published in 1977.

In the USSR, this field has developed since the late 1970s. The first work devoted to the theory of adaptive correction was published by V. P. Lukin in 1977. B. S. Agrovskii and V. V. Vorob'ev et al. (Institute of Atmospheric Physics RAS) and P. A. Konyaev (Institute of Atmospheric Optics SB RAS) studied AOS numerically. At the same time, M. A. Vorontsov, S. S. Chesnokov, V. A. Vysloukh, K. D. Egorov, and V. P. Kandidov (Moscow State University) published papers devoted to phase correction for nonlinear distortions. Special issues of the journal *Izvestiya Vysshikh Uchebnykh Zavedenii, Fizika* and monographs by M. A. Vorontsov and V. I. Shmal'gauzen (*Principles of Adaptive Optics*); V. P. Lukin (*Atmospheric Adaptive Optics*); and M. A. Vorontsov, A. V. Koryabin, and V. I. Shmal'gauzen (*Controllable Optical Systems*) review previous work in this field.

The current state of research on numerical modeling of adaptive optical systems can be characterized as follows. Basic numerical methods of solving the problems of optical wave propagation in randomly inhomogeneous media, including the thermal action of high-power beams, have been developed, and work on the development of numerical models of individual AOS components that takes into account their geometrical parameters and spatiotemporal resolution has been started, along with a search for the most efficient correction algorithms. This monograph summarizes the main results of our work in this field from 1985 to 1997.

The first chapter considers methodological aspects of numerical modeling of propagation of monochromatic coherent radiation in a randomly inhomogeneous, weakly absorbing medium. It describes numerical methods used to solve the inhomogeneous wave equation together with mathematical models of turbulent distortions and thermal inhomogeneities arising during optical radiation propagation through an absorbing medium. Numerical techniques of dynamic modeling of random phase screens are further developed and methods of modeling large-scale portions of the turbulence spectrum are described. In the last section of this chapter, the lens transformation is generalized to the case of an arbitrary optically inhomogeneous medium.

The second chapter describes numerical modeling of a closed AOS system and numerical models of a reference wave, sensors, and wavefront (WF) correctors. Mathematical models and the main points of numerical modeling are described for the following AOS components: an oncoming reference beam; natural and artificial reference stars; an ideal-square law and Hartmann wavefront sensors; and modal, segmented, and flexible adaptive mirrors.

In the third chapter, the problem of minimization and adaptive correction for turbulent distortions is solved. Here, the effect of the outer scale of turbulence on the main parameters of image formation in an atmosphere–telescope system, including the Strehl factor (SF) and the angular resolution (the width of the point spread function, PSF), is studied. The possibility of wavelength optimization is estimated quantitatively in a situation in which the size of the outer scale of turbulence is comparable to the aperture diameter. The angular resolution is further studied for incomplete (partial) correction for turbulent image distortions. In the last section of this chapter, the efficiency of phase correction is analyzed for extended paths and weak intensity fluctuations of the reference and corrected waves.

In the fourth chapter, the efficiency of adaptive correction for thermal activity is investigated. At the beginning of the chapter, the thermal effect of a wide-aperture high-power beam propagating along a vertical path represented by a composite nonlinear phase screen is analyzed. The parameters of beam power optimization and lower-mode correction for phase distortions are calculated for various intensity distributions over the beam cross section with allowance for the altitude dependence of the wind direction. The salient features of the functioning of phase-conjugation (PC) AOS used to correct for nonstationary action on a homogeneous horizontal path are further studied. A correlation between oscillations arising in these systems and phase dislocations in the reference wave is demonstrated. The results of numerical experiments for an AOS with the Hartmann WF sensor are given in the last section together with the modified phase conjugation algorithm and curves of power optimization that prove the efficiency of this modification.

The fifth chapter is devoted to an urgent problem of compensation for turbulent jitter in the image of an astronomical object when a laser guide star (LGS) is used as a reference source. Different configurations (bistatic and monostatic) of the system for measuring the random refraction are considered.

The efficiency of jitter correction is studied as a function of the ratio of the receiving and transmitting apertures. An algorithm of optimal correction for wavefront tilts is suggested and its efficiency is estimated.

The authors are indebted to their colleagues who were both formal and informal co-authors of the scientific results presented here: they include P.A. Konyaev, N.N. Maier, and E.V. Nosov; the staff of the Laboratory of Applied and Adaptive Optics; and many researchers at the Institute of Atmospheric Optics. Communications with them have helped determine the content of our monograph.

Vladimir P. Lukin
Boris V. Fortes

Adaptive Beaming and Imaging in the Turbulent Atmosphere

CHAPTER 1

Mathematical Simulation of Laser Beam Propagation in the Atmosphere

A key aspect of the numerical simulation of turbulent effects is generation of 2D random phase screens that imitate distortion of the wavefront during propagation through the atmosphere. One of the first papers devoted to numerical simulation of turbulent distortions of optical waves was by Buckley, who used a Fourier transform (method of spectral samples) to model 1D random phase screens [1]. In papers by Fleck, Morris, and Feit [2], Kandidov and Ledenev [3], and Konyaev [4], the method of spectral samples was used to generate 2D random phase screens in the problem of propagation of a coherent beam through a randomly inhomogeneous medium and, in particular, for estimating the efficiency of phase correction of atmospheric distortions. In a paper by Martin and Flatte [5], a similar method was used to study the probability density of intensity fluctuations.

The Fourier transform method was originally used for numerical simulation in radio engineering [6]. However, a prominent feature of the turbulent atmosphere as a randomly inhomogeneous medium is the wide range of spatial scales of refractive index inhomogeneities. To correctly simulate all of the scales (from inner to outer) of turbulent fluctuations, a computational grid should have at least a thousand nodes along every coordinate, which leads to huge computational time and expense.

To overcome the difficulties connected with the wide-band spectrum of atmospheric turbulence, it makes sense to use some "combined" method, which was first proposed in papers by Duncan and Collins [7, 8], as well as in a paper by Tel'pukhovskii and Chesnokov [9]. The main idea consists in the joint use of spectral (harmonic) and polynomial representations, each of which is used to simulate its own region of spatial scales: spectral decomposition is used to simulate small-scale inhomogeneities, and polynomial decomposition is used to represent scales larger than the size of the computational grid. This approach was further developed in Fortes and Lukin [10], where it was generalized to include nonstationary (dynamic) problems. More recent approaches for optical modeling [11] and physical modeling [12], as well as numerical approaches have also been developed [13-15].

In the following sections we apply our method for numerical solution to two tasks: high-power laser beam propagation in homogeneous media with absorption, and optical wave propagation through a random inhomogeneous turbulent atmosphere. As high-power coherent (laser) beams propagate through a nonturbid atmosphere, *thermal blooming* is one of the main factors causing distortion, along with turbulent fluctuations of the refractive index. This nonlinear effect has the lowest energy threshold and arises as a result of absorption of part of the beam energy and the formation of thermal inhomogeneity in the beam channel. The software developed by Konyaev [16] (Institute of Atmospheric Optics SB RAS, Tomsk, Russia) served as a basic model for numerical simulation of thermal blooming of a paraxial wave beam.

We have implemented several numerical schemes for solving differential equations that describe different hydrodynamic conditions of thermal blooming [17, 18]. In this chapter, we present examples that demonstrate the reliability of the results obtained.

1.1 Numerical Solution to Problems of Coherent Radiation Propagation

For both propagation of coherent beams and imaging in a randomly inhomogeneous medium, the wave equation for the electromagnetic field of an optical wave is the basis for a mathematical model. In the problems considered here, polarization effects are negligible, and the ratio of path length to aperture diameter is chosen so that the small-angle approximation (approximation of paraxial beams) is applicable for a scalar field amplitude [19-21].

1.1.1 Wave equation

Let us introduce a slowly varying component $E(\vec{\rho}, z, t)$ of the complex amplitude of an electromagnetic field in the following way:

$$\sqrt{\frac{cn_0}{8\pi}}\vec{E}(\vec{\rho}, z, t) = \vec{e}E(\vec{\rho}, z, t)\exp(ikz - i\omega t), \qquad (1.1.1)$$

so the intensity I is related to the component $E(\vec{\rho}, z, t)$ as

$$EE^* = I\,. \qquad (1.1.2)$$

Here, c is the speed of light in a vacuum, n_0 is the refractive index of a medium, $\vec{e}$ is the vector of polarization, $k = 2\pi/\lambda$ is wave number, ω is the frequency of electromagnetic oscillations, $\vec{\rho} = (x, y)$ is the vector of coordinates in the beam cross section (the beam is directed along the $0z$ axis), and t is time.

In the paraxial approximation, propagation of a monochromatic linearly polarized beam in a dielectrically inhomogeneous nonmagnetic medium is described by the parabolic equation for the complex amplitude E:

$$2ik\frac{\partial E}{\partial z} = \left[\frac{\partial^2}{\partial x^2} + \frac{\partial^2}{\partial y^2} + 2k^2(n^2/n_0^2 - 1)\right]E, \qquad (1.1.3)$$

or as

$$2ik\frac{\partial E}{\partial z} = \left(\frac{\partial^2}{\partial x^2} + \frac{\partial^2}{\partial y^2} + 2k^2\delta n\right)E, \qquad (1.1.4)$$

on the assumption that deviations of the refractive index from unity are small; i.e.,

$$n_0 \approx 1, \quad \delta n = (n-1) << 1. \qquad (1.1.5)$$

Initially, algorithms based on various finite-difference methods were used to solve the parabolic equation [22]. But, currently, the common method for solution in the domain of spatial frequencies of the complex amplitude E is the splitting algorithm applied together with a discrete Fourier transform (DFT).

The solution to the parabolic equation (1.1.4) corresponding to propagation of a wave from the plane z_l to the plane z_{l+1} can be written in operator form [2]:

$$E(x,y,z_{l+1}) = \exp\left[-\frac{i}{2k}\left(\Delta z\nabla_\perp^2 + 2k^2\int_{z_l}^{z_{l+1}}\delta n dz\right)\right]E(x,y,z_l), \qquad (1.1.6)$$

$$\nabla_\perp^2 = \frac{\partial^2}{\partial x^2} + \frac{\partial^2}{\partial y^2}.$$

This equation can be approximated [23, 24] by a symmetrized split operator:

$$E(x,y,z_{l+1}) = \hat{D}\left(\frac{1}{2}\Delta z\right)\hat{R}(z_l, z_{l+1})\hat{D}\left(\frac{1}{2}\Delta z\right)E(z_l) + O(\Delta z^2),$$

$$\hat{D}(\Delta z) = \exp\left(-i\frac{1}{2k}\Delta z\nabla_\perp^2\right), \quad R(z_l,\ z_{l+1}) = \exp\left(-ik\int_{z_l}^{z_{l+1}}\delta n dz\right). \qquad (1.1.7)$$

Here, the operator $\hat{R}(z_l,\ z_{l+1})$ describes *refraction* on inhomogeneities of the refractive index, and the operator $\hat{D}(\Delta z)$ corresponds to the solution of the problem of free *diffraction*. The second order of accuracy of this approximation has been proved analytically [2, 23] and confirmed by numerical experiments [24].

For optical waves, the *problem of free diffraction* at an arbitrary distance z can be solved using the representation for the complex amplitude in the form of a finite Fourier series [25]:

$$E(x,y,z)=\sum_{m=-N/2+1}^{N/2}\ \sum_{n=-N/2+1}^{N/2} E_{mn}(z)\exp\left[\frac{2\pi i}{L}(xm+yn)\right], \quad (1.1.8)$$

where

$$E_{mn}(z)=\frac{1}{4\pi^2L^2}\int_0^L\!\!\int_0^L dxdyE(x,y,z)\exp\left[-i\frac{2\pi}{L}(xm+yn)\right] \quad (1.1.9)$$

are the expansion coefficients, L is the size of the domain of expansion, and N is the number of terms in the series. It is also assumed that the spectrum of spatial frequencies for the function $E(x, y, z)$ is finite and the function itself is periodic or can be supplemented by a periodic function. In a numerical simulation, a continuous field $E(x, y, z)$ is replaced by a discrete field defined at the nodes of a computational grid. The transition from the domain of the original function to the spectral space and back is performed by DFT.

Substituting the spectral representation into the parabolic equation (1.1.4), we obtain

$$2ik\frac{\partial E_{mn}}{\partial z}=-\frac{4\pi^2}{L^2}(m^2+n^2)E_{mn} \quad (1.1.10)$$

with the following exact solution

$$E_{mn}(z)=E_{mn}(z=0)\exp\left[-\frac{4\pi^2 z}{2ikL^2}(m^2+n^2)\right]. \quad (1.1.11)$$

To solve *the problem of refraction* in the layer Δz, we need to obtain the numerical representation for inhomogeneities $\delta n(\vec{\rho}, z)$ of the refractive index. Refraction is described as beam passage through a phase screen:

$$\hat{R}(z_l, z_{l+1})=\exp[i\varphi_l(\vec{\rho})], \qquad \varphi_l(\vec{\rho})=-ik\int_{z_l}^{z_{l+1}}\delta n(\vec{\rho},z)\,dz. \quad (1.1.12)$$

The mathematical model of refractive index inhomogeneities depends on the process by which they arise. Here we consider two effects: a lowest-threshold nonlinear effect known as random thermal blooming and fluctuations induced by atmospheric turbulence.

1.1.2 Thermal blooming of high-power laser beams

The longitudinal scale of variability for thermal inhomogeneities induced in the propagation channel of a high-power laser beam is comparable to the diffraction length of the beam. In the interval Δz, the equation for the phase screen can be approximated by the product of a step length Δz and the refractive index distribution at the center of the interval $[z_l; z_l+\Delta z]$:

$$\varphi_l(\vec{\rho}) = k\Delta z \delta n\left(\vec{\rho}, z_l + \frac{1}{2}\Delta z\right) + O(\Delta z^2) . \tag{1.1.13}$$

It follows from the above that we have only to determine perturbations of the refractive index in some planes, the positions of which are determined by the scheme of the splitting algorithm.

Heating of the medium that is caused by absorption of radiation energy induces variation of its density, which leads to a decrease in the refractive index related to the density ρ by the following law [26]:

$$\delta n = K\rho , \tag{1.1.14}$$

where K is a constant equal to two-thirds of the polarization factor of a molecule or gas atom.

In the isobaric approximation, the density of the medium is explicitly related to temperature by the ideal gas law, so variations of the refractive index can be expressed through temperature variations:

$$\delta n \approx \frac{\partial n}{\partial T}(T - T_0) = n'_T \delta T. \tag{1.1.15}$$

The isobaric approximation is valid for the normal atmospheric conditions. Exceptions are fast scanning of a continuous-wave (cw) high-power beam when the beam speed with respect to the medium is greater than the sonic speed, and when the pulse duration τ_p is comparable with the acoustic time τ_s:

$$\tau_p = \tau_S = a / c_S , \tag{1.1.16}$$

where a is the beam size and c_s is the sonic speed.

When the isobaric approximation is valid, the distribution of the refractive index in the beam cross section is determined by the heat balance, which is described by the heat transfer equation for the temperature field $T(x, y, z)$:

$$\frac{\partial T(x, y, z)}{\partial t} + \vec{V}_\perp \nabla T - \chi \Delta_\perp T = \frac{\alpha}{\rho_0 C_p} I , \tag{1.1.17}$$

where $\vec{V}_\perp = (V_x, V_y)$ is the transverse component of the beam velocity relative to the medium, χ is heat conductivity, ρ_0 is the specific density of the medium, α is the absorption coefficient, and C_p is the specific heat at constant.

When the isobaric approximation becomes invalid, variations in the density of the medium are described by the linearized equations of hydrodynamics, which follow from the law of continuity and the laws of impetus and energy conservation [21, 26]:

$$\frac{d\rho_1}{dt} + \rho_0 \nabla \mathrm{v}_1 = 0, \quad \frac{d}{dt} = \frac{\partial}{\partial t} + \mathrm{v}_0 \frac{\partial}{\partial x}, \tag{1.1.18}$$

$$\rho_0 \frac{d\mathrm{v}_1}{dt} = -\nabla p_1, \tag{1.1.19}$$

$$\frac{d}{dt}(p_1 - c_S^2 \rho_1) = (\gamma - 1)\alpha I. \tag{1.1.20}$$

These equations are valid for small perturbations of density ρ_1, pressure p_1, and local speed of the medium flow v_1 with respect to the unperturbed values of density ρ_0, pressure p_0, and local speed of medium flow v_0 along the x-axis. By eliminating the variables describing perturbation of speed and pressure, Eqs. (1.1.18)–(1.1.20) can be transformed into the following equation, which describes perturbations of the density:

$$\left(\frac{d^2}{dt^2} - c_S^2 \nabla^2 \right) \frac{d\rho_1}{dt} = (\gamma - 1)\alpha \nabla^2 I. \tag{1.1.21}$$

For a steady-state condition, this equation takes the form

$$\left[\frac{\partial^2}{\partial y^2} + (1 - M^2) \frac{\partial^2}{\partial x^2} \right] \frac{d\rho_1}{dx} = \frac{(\gamma - 1)}{c_S^2 \mathrm{v}} \alpha \nabla^2 I, \tag{1.1.22}$$

where $M = \mathrm{v}/c_S$ is Mach's number and $\gamma = C_p/C_v$ is the specific heat ratio at a constant volume.

When the flow rate of the medium is low ($M = \mathrm{v}/c_S \ll 1$), Eq. (1.1.22) transforms into the equation written in the isobaric approximation:

$$\frac{d\rho_1}{dx} = \frac{(\gamma - 1)}{c_S^2 \mathrm{v}} \alpha I. \tag{1.1.23}$$

Solution of this equation yields a result equivalent to that of Eq. (1.1.17), with $v_y = 0$ and $\chi = 0$.

With no wind, under *conditions of gravitational convection*, the heat balance equation should be solved by a set of equations for incompressible liquid hydrodynamics. The plane (2D) flow of this liquid in the Boussinesq–Oberbeck approximation is described by the vortex (vorticity) function ω and the stream function ψ [27] equations:

$$\frac{\partial \omega}{\partial t} + \vec{v}\vec{\nabla}_{\perp}\omega = \nu\Delta_{\perp}\omega + \beta g\frac{\partial T}{\partial x}\ ; \tag{1.1.24}$$

$$\Delta_{\perp}\psi = -\omega, \tag{1.1.25}$$

where β is the volume expansion coefficient, ν is the kinematic viscosity coefficient, and g is the absolute value of the acceleration of gravity directed along the 0y axis.

The local speed of the flow in the equations for temperature and vorticity is now a function of transverse coordinates. Its components are related to the partial derivatives of the stream function as follows:

$$v_x = \frac{\partial \psi}{\partial y}, \quad v_y = -\frac{\partial \psi}{\partial x}\ . \tag{1.1.26}$$

When describing thermal blooming under conditions of free convection, Eqs. (1.1.17), (1.1.24), and (1.1.25) are complemented by the corresponding initial and boundary conditions. In the case of propagation in a closed space, they are the conditions of adhesion and zero flow speed at the boundary Γ for the normal and tangential components of the flow velocity:

$$v_n(x, y) = 0, \quad v_s(x, y) = 0, \quad (x, y) \in \Gamma\ . \tag{1.1.27}$$

These conditions in turn determine the boundary conditions for the stream function and its derivative with respect to the normal to the boundary surface:

$$\psi(x, y) = 0, \quad \frac{\partial \psi}{\partial n} = 0, \quad (x, y) \in \Gamma. \tag{1.1.28}$$

The boundary and initial conditions for the temperature field are usually set to be zero.

So the mathematical model of thermal blooming of high-power coherent laser beams in a low-absorbing medium includes the parabolic wave equation for the scalar complex amplitude [Eq. (1.1.4)] and the corresponding material equation (or set of equations) describing density and temperature variations and determining the distribution of the refractive index in the beam channel:

$$\begin{cases} 2ik\dfrac{\partial E}{\partial z} = \left(\dfrac{\partial^2}{\partial x^2} + \dfrac{\partial^2}{\partial y^2} + 2k^2\delta n\right)E \\ \hat{M}(\delta n) = \alpha I \end{cases}, \qquad (1.1.29)$$

where the operator $\hat{M}$ describes the relation between the induced optical inhomogeneities and the absorbed energy αI.

Below we present the results of applying our numerical technique for estimating the thermal blooming distortions of coherent beams propagating along atmospheric paths that were developed in Refs. [4], [16-18], [24], [30], and [49]. To allow for the regular altitudinal variation of thermodynamic parameters, we have used statistical seasonal atmospheric models, built from refined data obtained by the Institute of Atmospheric Optics in Tomsk [56-58]. The altitude profile of the molecular absorption coefficient for summer and winter mean-latitude models was obtained by means of a software program that calculates molecular absorption line by line [59].

As an example of implementation of the mathematical model, let us consider the results computed for thermal blooming of a focused Gaussian beam crossing a thin layer L of a nonlinear medium (nonlinear phase screen). In this example, the approximation of a nonlinear phase screen indicates that the thickness L of the layer is much shorter than the beam focal length and the diffraction length $L_d = ka_0^2$ of the beam and that the Bouguer extinction is low ($\alpha L \ll 1$).

In Fig. 1.1.1, the dynamics of thermal blooming under conditions of forced convection are illustrated for a convective flow speed that is much lower than the sonic speed [isobaric approximation, Eq. (1.1.17)] and for the heat conductivity $\chi = 0$. In this case, the sole parameter of the problem [56-59] is the integral nonlinearity of the medium layer:

$$P_N = \frac{2\pi}{\lambda}\frac{\alpha a_0 I_0}{\rho C_p V_\perp} n_T' L,$$

where I_0 is the initial intensity of the beam, a_0 is the initial size of the Gaussian beam, and α is the Bouguer extinction of media. When transient processes end, the phase screen can be described as an integral of the normalized beam intensity:

$$\varphi(x, y) = P_N \int_{-\infty}^{x} I(\xi, y) \Big/ I_0 d\xi = P_N \int_{-\infty}^{x} \exp\left(-\xi^2 - y^2\right) d\xi. \qquad (1.1.30)$$

At a large distance x, the value of phase φ at the $0z$ axis ($y = 0$) tends toward $\sqrt{\pi}P_N \approx 1.77P_N$. In calculations, the value of P_N was assumed to be equal to 10; this value corresponds to the maximal phase difference of 17.7 rad or 2.8λ. For such nonlinear distortions, the steady-state maximal value of the focus intensity decreases more than 10 times. Figure 1.1.1 shows two examples of calculations for different directions of the wind vector. For every instant of time t (here, time t is normalized at a value of $a_0/V_\perp$, where $V_\perp$ is wind velocity) the 2D distributions of phase distortions in the plane of the emitting aperture are shown, along with the intensity distribution in the focal plane for the Gaussian beam. In the first case the wind was directed along the ordinate; and in the second case, the angle between the wind and the ordinate was 45 deg.

Figure 1.1.2 shows typical images of thermal blooming of laser radiation under different stream velocities. The numbers in each image box of Figs. 1.1.1 and 1.1.2 show the maximum and minimum values of the functions $I(\vec{\kappa})$ and $\varphi(\vec{\rho})$. It can be seen that the estimate of phase distortions in the steady state (17.7 rad) is in good agreement with the results obtained in the numerical experiment (16.9 and 16 rad). The difference of 5–10% can be attributed to the error of the numerical model and the fact that the transient processes did not end completely.

The calculations have been performed by the monotonic conservative procedure with the second-order approximation, differences were taken with respect to the stream [27], and the model viscosity (heat conductivity) was compensated for according to Samarskii's algorithm [28]. The advantage of this method is that it makes it possible to solve the dynamic problem of heat transfer at an arbitrary orientation of the wind vector. The method is stable for both small and large intervals of time discretization and it allows the solution of the heat transfer problem when wind velocity depends on the transverse coordinates (x,y). Thus, the calculations can be performed for the problem of thermal blooming under free (gravitational) convection conditions [Eq. (1.1.17) together with Eqs. (1.1.24) and (1.1.25)], as well as the boundary conditions of adhesion and zero speed. Some sample calculations are given in Ref. [29]. In this text we do not consider these convection conditions in greater detail because gravitational convection is atypical for open atmospheric paths.

Next we consider the situation of a high-power beam with scanning. In this case, the speed of beam transition with respect to the medium at some distance from the source can be close to or even higher than sonic speed. In Fig. 1.1.2, density perturbations are shown for forced convection conditions with the speed of the flow close to sonic speed [Eq. (1.1.22)]. The value of the parameter P_N was assumed to be two times smaller than in the previous example. In the isobaric approximation, this corresponds to the maximal phase change on the path (8.8 rad). It is seen that for a Mach number range of $M = 0.5$–0.7 the results do not differ widely. With a further increase in M (i.e., M approaches unity), phase distortions increase sharply. For an M greater than unity, we obtain a solution approximately equal to the results of the isobaric approximation, but two times greater. It can be shown analytically that for M approaching infinity this conclusion is correct.

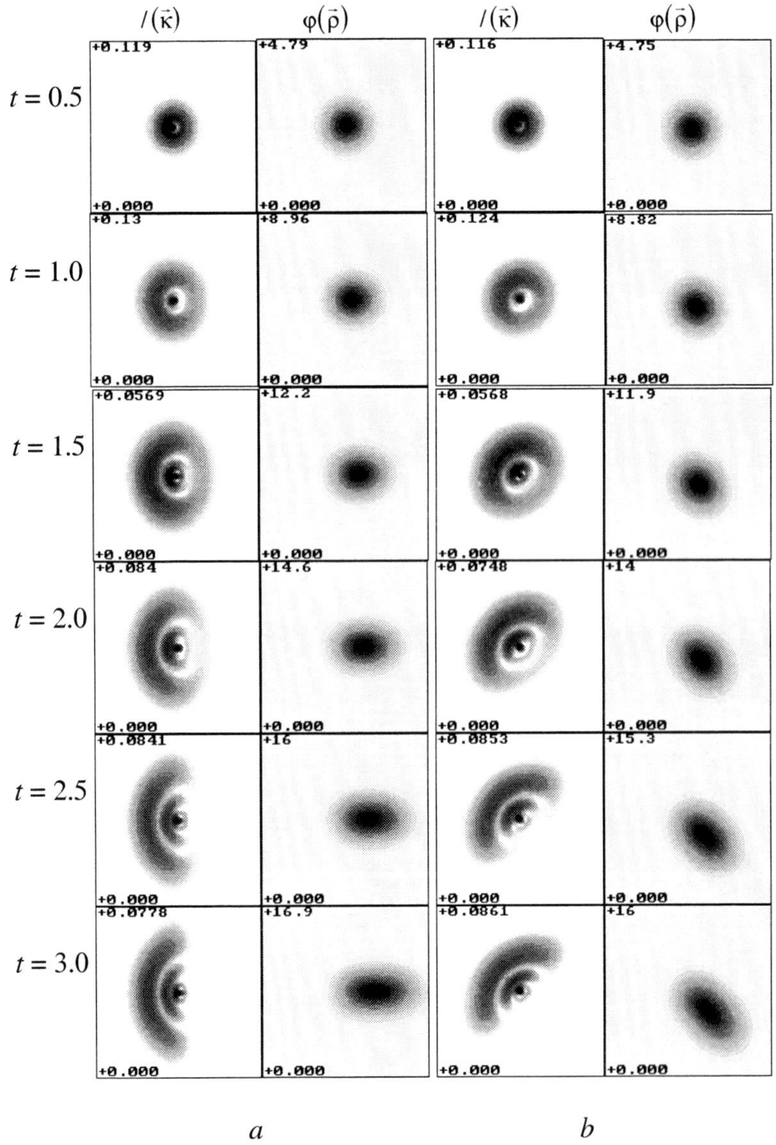

Figure 1.1.1. Dynamics of thermal blooming under conditions of forced convection (isobaric approximation): (a) wind is directed along the 0*x*-axis and (*b*) at an angle of 45 deg to the 0*x*-axis; $I(\vec{\kappa})$ is the intensity distribution in the far zone; $\varphi(\vec{\rho})$ is the phase screen.

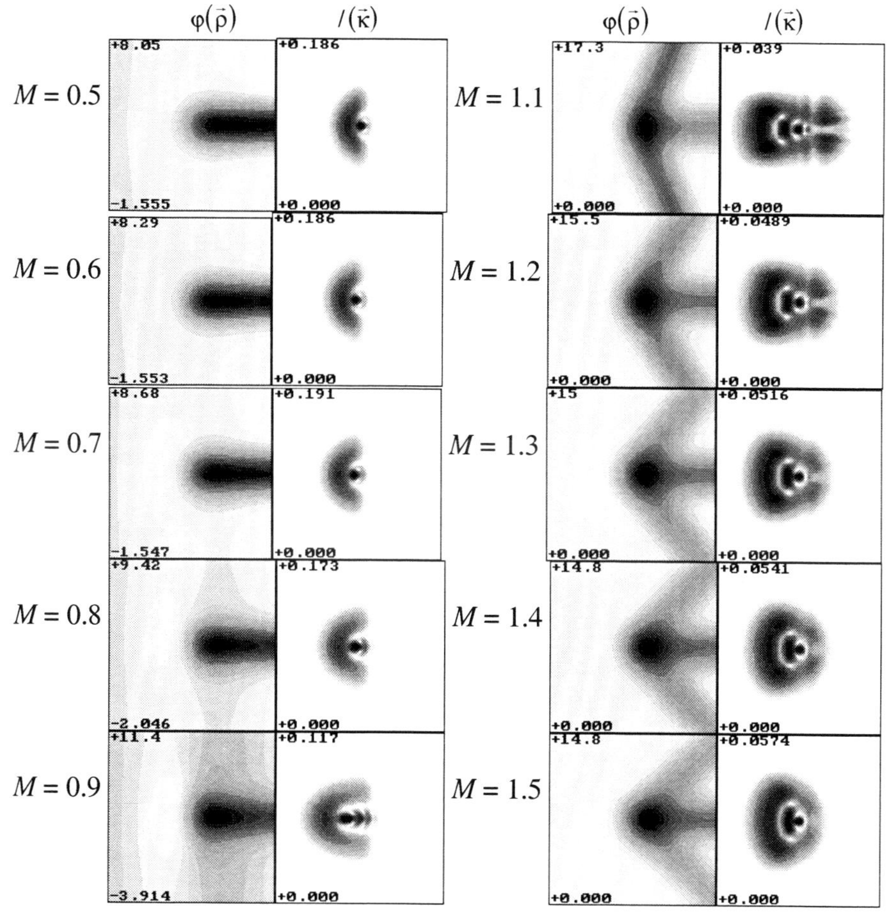

Figure 1.1.2. Thermal blooming behind the phase screen under conditions of forced convection (stationary solution of the linearized equations of hydrodynamics); $I(\vec{\kappa})$ is the intensity distribution in the far zone, $\varphi(\vec{\rho})$ is the phase screen.

1.1.3 Turbulent distortions of a wavefront

Under conditions of turbulent fluctuation, the longitudinal scale of variability of the refractive index is on the same order as the inner scale of turbulence l_0, which is usually much less than the discretization step Δz for a reasonable (in terms of computational expense) number of integration steps for the wave equation. In this case, integration of refractive index inhomogeneities along the z coordinate should be performed analytically. And, because a statistical method is used to describe turbulent fluctuations $\delta n(\vec{\rho}, z)$, the integration yields the equation for statistical characteristics, for example, the correlation function of phase fluctuations:

$$B_\varphi(\vec\rho) = \langle \varphi(\vec r + \vec\rho)\varphi(\vec r)\rangle = k^2 \int^{\Delta z}\int^{\Delta z} dz' dz'' \langle n(\vec r + \vec\rho,\ z_l + z')n(\vec r + \vec\rho,\ z_l + z'')\rangle . \tag{1.1.31}$$

The corresponding spectral density F_φ obtained in Ref. [2] for the von Karman spectrum of turbulence has the form

$$F_\varphi(\vec\kappa) = 2\pi k^2 \Delta z \Phi_n(\vec\kappa_\perp, \kappa_z = 0, L_0) K\left(\vec\kappa, \frac{L_0}{\Delta z}\right), \tag{1.1.32}$$

where Φ_n is the 3D spectral density of refractive index fluctuations, L_0 is the outer scale, K is the correction function approaching unity for small values of the second argument (which corresponds to the Markov approximation), and $\vec\kappa$ is the spatial frequency.

When considering propagation of high-power laser beams in a randomly inhomogeneous medium, thermal distortions caused by heating of the medium as a result of radiation propagation should be taken into account, along with turbulent fluctuations of temperature. Within the integration step along the longitudinal coordinate, turbulent and nonlinear phase distortions can be considered as additive; i.e.,

$$\varphi(\vec\rho, t) = k \int_{z_l}^{z_{l+1}} \delta n(\vec\rho, z, t) dz \approx \varphi_1(\vec\rho, t) + \varphi_2(\vec\rho, t), \tag{1.1.33}$$

where φ_1 stands for nonlinear distortions and φ_2 stands for turbulent ones.

Thus, during numerical simulation of optical wave propagation through a randomly inhomogeneous medium, the solution to the problem of refraction at every step of integration of the differential wave equation by the splitting algorithm is reduced to generation of random phase screens that meet the preset statistics. Simulation of random processes is a special branch of computational mathematics and its application to the study of turbulent distortions of optical waves in the atmosphere will be considered later. Here we consider only such problems as the choice of discretization step along the transverse coordinate $\Delta\rho$ and the size of the averaging sample.

The applicability of this numerical model is determined mainly by its ability to represent the complex amplitude of an optical wave as a discrete function. According to sampling theorem, this condition limits the spatial width $\kappa_{\max}$ of the spectrum of this function as

$$\kappa_{\max} < \pi / \Delta\rho , \tag{1.1.34}$$

where $\Delta\rho$ is a spatial step.

The width of the spectrum can be estimated using the well-known fact that the angular size of the image of a point object W observed through a turbulent atmosphere is approximately equal to the ratio of the wavelength λ to the coherence length r_0:

$$W \approx \lambda / r_0, \qquad r_0^{-5/3} = 0.423k^2 \int C_n^2 dz\,. \tag{1.1.35}$$

Since a lens—the simplest imaging system—in effect performs Fourier transformation of the incident field, the size of W corresponds to the width of the spatial spectrum; i.e., $\kappa_{\max} \approx W / \lambda$. It should be noted that the ratio λ / r_0 corresponds to the image of a full-width half-maximum (FWHM). The radius of the spot including 95% of the energy is three times larger. Correspondingly, for a plane wave that has passed through a layer of atmospheric turbulence, the width of the spectrum can be estimated as

$$\kappa_{\max} \approx 3W / \lambda = 3 / r_0\,. \tag{1.1.36}$$

This leads to the following restriction for the discretization step:

$$\Delta\rho < r_0 \frac{\pi}{3} - 1\,. \tag{1.1.37}$$

Another restriction occurs because small-scale turbulent distortions with the spatial frequencies $\kappa > \kappa_{\max}$ are lost during discretization with step $\Delta\rho$. This leads to underestimation of the phase distortion variance σ^2 by a value approximately equal to the integral over the corresponding spectral interval. Let us impose a limit on this value corresponding to the "λ/10" criterion:

$$\begin{aligned} \sigma^2 &= 2\pi \int_{\kappa_{\max}}^{\infty} F_\varphi(\kappa)\kappa d\kappa = 2\pi 0.489 r_0^{-5/3} \int_{\kappa_{\max}}^{\infty} \kappa^{-8/3} d\kappa \\ &= 0.489 r_0^{-5/3} \frac{3}{5} \kappa_{\max}^{-5/3} = 0.3[\Delta\rho / (\pi r_0)]^{5/3} < (0.2\pi)^2 \approx 0.4. \end{aligned} \tag{1.1.38}$$

Here we have a less strict condition for the discretization step; namely, $\Delta\rho < \pi r_0$.

Let us now consider a sample size of N_{smp}. For the problems considered here, we are interested only in the moment of the first order, i.e., the mean intensity. The variance of the estimation errors for a mean value is related to the fluctuation variance as follows:

$$\sigma^2_{\text{estm err}} = \frac{\sigma_I^2}{N_{\text{smp}}}\,. \tag{1.1.39}$$

From this equation we obtain the following condition, which ensures the 10% root-mean-square (rms) error of estimation:

$$N_{\text{smp}} = \frac{\sigma_I^2}{\sigma_{\text{estm err}}^2} = \frac{\sigma_I^2}{(0.1\langle I \rangle)^2} = 100\frac{\sigma_I^2}{\langle I \rangle^2}. \tag{1.1.40}$$

Now we can estimate the sample size a priori using the known relative variance of intensity fluctuations and the well-known effect of saturation of intensity fluctuations. Experimental data [21] and the results of numerical experiments [5, 30-32] indicate that the normalized variance of intensity fluctuations almost never exceeds unity. This is true both for a plane wave and for close-to-axis regions of a spatially limited beam or in the image plane. So, we have $N_{\text{smp}} = 100$ for a 10% estimation accuracy of the mean intensity at the optical axis of the system.

A less strict condition is typical for estimating the efficiency of adaptive correction because even incomplete correction of distortions significantly suppresses intensity fluctuations. In the calculations whose results are given here, the sample size $N_{\text{smp}} = 100$ allowed the mean intensity of the corrected beam or image to be estimated with a 5% accuracy.

Thus, we have determined the requirements for the basic parameters of numerical simulation: the step of the computational grid and the size of the averaging sample. The issues of sample generation are considered in the following sections.

1.2 Generation of 2D Random Phase Screens by the Fourier Transform Method

In this section we consider the problem of generating 2D random phase screens on an equidistant computational grid. This problem falls in the class of problems of numerical simulation of normal random processes with known statistics. When describing optical wave propagation in a turbulent atmosphere, the starting point of the simulation is the "2/3" law [33-34] for the structure function of refractive index fluctuations:

$$D_n(\rho) = C_n^2 \rho^{2/3}, \tag{1.2.1}$$

where C_n^2 is the structure constant of refractive index fluctuations.

The equation for the structure function D_n determines the form of the 3D spectral density:

$$\Phi_n(\vec{\kappa}) = 0.033\ C_n^2\ \kappa^{-11/3}, \quad \kappa = \left(\kappa_x^2 + \kappa_y^2 + \kappa_z^2\right)^{1/2}. \tag{1.2.2}$$

If it is necessary to take into account the effects connected with the inner l_0 or outer L_0 scales of turbulence, various modifications of the turbulent spectrum are used [35]; for example,

$$\Phi_n(\kappa) = 0.033\, C_n^2\, (\kappa^2 + \kappa_0^2)^{-11/6} \exp(-\kappa^2/\kappa_m^2);$$
$$\kappa_0 = 2\pi/L_0,\ \kappa_m = 2\pi/l_0. \qquad (1.2.3)$$

For a plane wave passed through a layer of a randomly inhomogeneous medium with length L, the 2D spectral density of phase fluctuations in the geometric optics approximation has the form

$$F_S(\kappa) \approx 2\pi k^2 L \Phi_n(\kappa), \qquad (1.2.4)$$

where $k = 2\pi/\lambda$ is the wave number of the optical radiation. This equation is derived in the Markovian approximation corresponding to the delta correlation of refractive index fluctuations along the direction of propagation.

Let us consider simulation of a stationary random process with a given spectral density $F(\kappa)$. In a description of random processes, the process S is usually represented as the stochastic Fourier–Stieltjes integral [36]:

$$S(\vec{\rho}) = \iint \exp(i\vec{\kappa}\vec{\rho})\ d^2H(\vec{\kappa}). \qquad (1.2.5)$$

In stationary random processes, spectral components $d^2H(\vec{\kappa})$ of random processes are delta correlated and related to the spectral density by the following equation:

$$\left\langle d^2H(\vec{\kappa})d^2H^*(\vec{\kappa}')\right\rangle = F_S(\vec{\kappa})\delta(\vec{\kappa}-\vec{\kappa}')d^2\kappa d^2\kappa'. \qquad (1.2.6)$$

Hereafter, angular brackets indicate averaging over an ensemble, and an asterisk (*) is used for complex conjugation.

In numerical simulation of stationary random processes, the field $S(\vec{\rho})$ discretized over spatial variables should be generated in such a way that the statistical and correlation properties of the discrete random processes correspond to the properties of the initial continuous process. According to the Fourier transform method, the simulated process is represented as a truncated Fourier series whose coefficients form the 2D $N \times N$ matrix of spectral counts $\tilde{S}$:

$$S(\vec{\rho}) = \sum_{-N/2}^{N/2} \sum_{-N/2}^{N/2} \tilde{S}(\vec{\kappa}_{L,M}) \exp(i\vec{\kappa}_{L,M}\vec{\rho}), \qquad (1.2.7)$$

where L and M are summation indices. In numerical experiments, we are usually interested in values of a function at the nodes of the equidistant computational grid,

$$\vec{\rho}_{I,J} = \vec{e}_x I\Delta\rho + \vec{e}_y J\Delta\rho, \quad I,\ J = 0,\ 1,\ ...,\ N; \tag{1.2.8}$$

so the double sum in Eq. (1.2.7) can be rewritten as

$$S(\vec{\rho}_{I,J}) = S_{I,J} = \sum_{-N/2}^{N/2}\sum_{-N/2}^{N/2} \tilde{S}_{L,M} \exp[i(I\times L + J\times M)\Delta\kappa\Delta\rho]. \tag{1.2.9}$$

Assuming that

$$\Delta\kappa = \frac{2\pi}{N\Delta\rho} \tag{1.2.10}$$

and changing the summation limits, we obtain the standard equation of a 2D DFT:

$$S_{I,J} = \sum_{0}^{N-1}\sum_{0}^{N-1} \tilde{S}_{L,M} \exp\left[\frac{2\pi i}{N}(I\times L + J\times M)\right], \tag{1.2.11}$$

or an equivalent formula,

$$S_{I,J} = \sum_{-N/2}^{N-1} \exp\left(\frac{2\pi i}{N} J\times M\right) \sum_{-N/2}^{N-1} \tilde{S}_{L,M} \exp\left(\frac{2\pi i}{N} I\times L\right) \tag{1.2.12}$$

It has been shown [1, 6] that to obtain the given correlation properties, the coefficients of the Fourier series should be related to the spectral density of the initial process by the following equation:

$$\left\langle \left|\tilde{S}_{I,J}\right|^2 \right\rangle = F_S(\kappa_{LM})\Delta\kappa^2, \quad \kappa_{LM} = \Delta\kappa\sqrt{L^2 + M^2}, \tag{1.2.13}$$

where $\Delta\kappa$ is the discretization step for the spatial frequency. In this case, the correlation function of the series obtained corresponds to the correlation function of the simulated random process if the correlation length of the random process and the values used for the correlation function are smaller than the size of the spatial region covered by the computational grid. This restriction is due to the periodicity of the series resulting from the DFT.

When selecting the form of the distribution function for the probability density of random Fourier coefficients, we have a certain amount of freedom. For example, it has been shown [37] that both the normal and homogeneous distributions of counts in a spectral domain can be used to obtain the series of correlated Gaussian numbers. In the latter case, the multidimensional characteristic function of the correlated series obtained converges to the multidimensional characteristic function of the Gaussian process as $O(1/\sqrt{N})$. To obtain a random sample of spectral counts, other authors have used the χ^2-distribution for the absolute values of complex counts and the homogeneous (in interval [0, 2π]) distribution for the arguments (this is equivalent to the normal distribution of their real and imaginary parts). Indeed, the real and imaginary parts of counts obtained as a result of a discrete Fourier transform applied to a correlated series of normal random numbers are also normally distributed random numbers. From this point of view, it is quite justifiable to use the Gaussian function.

A shortcoming that is common to all of the references quoted above is the absence of an explanation of why a particular form of the distribution function is chosen for random counts from the viewpoint of simulation of a steady-state random process. However, usually the problem is one of obtaining a series of correlated numbers rather than the simulation of a random process. The arguments given below clearly indicate that to make the procedure more rigorous from a mathematical point of view, one should take the constant absolute value and the uniformly distributed argument for spectral counts in the interval [0; 2π].

Rewrite the spectral representation in Eq. (1.2.5) as an ordinary Fourier integral:

$$S(\vec{\rho}) = \iint \exp(i\vec{\kappa}\vec{\rho})\, d^2H_S(\vec{\kappa}) = \iint A(\vec{\kappa}) \exp(i\vec{\kappa}\vec{\rho})\, d^2\kappa, \qquad (1.2.14)$$

keeping in mind that the spectral amplitude

$$A(\vec{\kappa}) = \frac{1}{(2\pi)^2} \iint S(\vec{\rho}) \exp(-i\vec{\kappa}\vec{\rho})\, d^2\kappa \qquad (1.2.15)$$

is a generalized random function. Since $H(\vec{\kappa})$ is a function with independent random increments $a(\vec{\kappa})$ [36],

$$a(\vec{\kappa}) = H(\vec{\kappa} + d\vec{\kappa}) - H(\vec{\kappa}), \qquad (1.2.16)$$

the spectral amplitude must include the delta function

$$A(\vec{\kappa}) = a(\vec{\kappa})\, \delta^{1/2}(\vec{\kappa}). \qquad (1.2.17)$$

Taking this into account and rewriting Eq. (1.2.6) as

$$\left\langle A(\vec{\kappa})A^*(\vec{\kappa}')\right\rangle = F_S(\vec{\kappa})\,\delta(\vec{\kappa}-\vec{\kappa}'), \qquad (1.2.18)$$

we can see that the density variance of the spectral amplitude is

$$\left\langle A(\vec{\kappa})A^*(\vec{\kappa})\right\rangle = \delta(\vec{0})\,F_S(\vec{\kappa}), \qquad (1.2.19)$$

and the variance of $a(\vec{\kappa})$ is equal to the spectral density

$$\left\langle a(\vec{\kappa})a^*(\vec{\kappa})\right\rangle = F_S(\vec{\kappa}). \qquad (1.2.20)$$

Thus, the spectral amplitude A should be considered as a delta-correlated generalized random function. Therefore, the Fourier integral in Eq. (1.2.14) cannot be estimated according to the trapezium or rectangle rule, as was done in Ref. [2]. Instead, we should first construct the spectral approximation of the random process S. The initial random process is represented as a sum of narrow-band processes,

$$S(\vec{\rho}) = \sum_{n=1}^{N} S_n(\vec{\rho}), \qquad (1.2.21)$$

each characterized by the spectrum located in a narrow-band $\Delta\kappa$:

$$F_n(\vec{\kappa}) = \begin{cases} F(\vec{\kappa}), \vec{\kappa} \in [\vec{\kappa}_n, \vec{\kappa}_n + \Delta\vec{\kappa}] \\ 0, \quad \vec{\kappa} \notin [\vec{\kappa}_n, \vec{\kappa}_n + \Delta\vec{\kappa}] \end{cases}, \qquad (1.2.22)$$

and connected by the Fourier transform with the corresponding generalized spectral amplitude A_n:

$$S_n(\vec{\rho}) = \int A_n(\vec{\rho}) \exp(i\vec{\kappa}\vec{\rho})\, d^2\rho\,; \qquad (1.2.23)$$

$$\left\langle A_n(\vec{\kappa})A_n^*(\vec{\kappa}')\right\rangle = \delta(\vec{\kappa}-\vec{\kappa}')\,F_n(\vec{\kappa}). \qquad (1.2.24)$$

When moving from the integral to the Fourier series, we replace the sum of narrow-band random processes with the sum of harmonics with random coefficients. At this point, we face the problem of selecting the probability distribution function for the coefficients of the series. Since our purpose is to obtain a random process in which the given spectral density reflects the distribution of fluctuation "energy" over the spatial frequencies, the normalized energy of every narrow-band process must coincide with the normalized variance of the corresponding harmonic of the Fourier series:

$$\frac{\left|\tilde{S}_{L,M}\right|^2}{\sum\limits_{L,M}\left\langle\left|\tilde{S}_{L,M}\right|^2\right\rangle} = \frac{\int\limits_{\vec{\kappa}_{L,M}-\Delta\kappa/2}^{\vec{\kappa}_{L,M}+\Delta\kappa/2} \left|A(\vec{\kappa})\right|^2 d^2\kappa}{\int\limits_{-\infty}^{+\infty}\left\langle\left|A(\vec{\kappa})\right|^2\right\rangle d^2\kappa}. \tag{1.2.25}$$

The denominators here contain the energy of fluctuations in the entire frequency band.

Consider the integral in the numerator on the right-hand side of the equation. If the spectral density varies insignificantly within the bandwidth $\Delta\vec{\kappa}$, then the density of the spectral amplitude $A(\vec{\kappa})$ and the square of its absolute value have the same statistical characteristics throughout the integration interval. Let us show that integration over the band $\Delta\vec{\kappa}$ can be considered as averaging over an ensemble. For this purpose, use the definition of the integral as a limit of the integral sum used:

$$\int\limits_{\vec{\kappa}_{L,M}-\Delta\kappa/2}^{\vec{\kappa}_{L,M}+\Delta\kappa/2} \left|A(\vec{\kappa})\right|^2 d^2\kappa = \lim_{N_\kappa\to\infty}\left(\frac{\Delta\kappa}{N_\kappa}\right)^2 \sum_{I,J=1}^{N_\kappa}\left|A(\vec{\kappa}_{I,J})\right|^2$$

$$\approx \Delta\kappa^2 \lim_{N_\kappa\to\infty}\left(\frac{1}{N_\kappa}\right)^2 \sum_{I,J=1}^{N_\kappa}\left|A(\vec{\kappa}_{L,M})\right|^2 = \left\langle\left|A(\vec{\kappa}_{L,M})\right|^2\right\rangle\Delta\kappa^2 = F(\vec{\kappa}_{L,M})\Delta\kappa^2\delta(\vec{0}). \tag{1.2.26}$$

Here, the parameter N_κ refers to the number of division points in the interval $\Delta\kappa$; $\kappa_{I,J}$ are the points at which the value of the function is selected when calculating the integral sum. Here we consider $A(\kappa_{I,J})$ as terms of the N_κ-sized sample of a random parameter $A(\kappa_{L,M})$, where $\kappa_{L,M}$ is a point inside the interval $\Delta\kappa$. This approximation is valid if the integral $\Delta\kappa$ is small and the statistics of the random field $A(\kappa)$ vary insignificantly within its limits.

Calculation of the integral in the numerator of the right-hand side of Eq. (1.2.23) gives

$$\int\limits_{-\infty}^{+\infty}\left\langle\left|A(\vec{\kappa})\right|^2\right\rangle d^2\kappa = \int\limits_{-\infty}^{+\infty} F(\vec{\kappa})\delta(\vec{0})\, d^2\kappa = \sigma^2\delta(\vec{0}). \tag{1.2.27}$$

The sum in the numerator on the left-hand side of Eq. (1.2.25) is the variance of the process with a discrete spectrum. If we require that the variances of both processes be equal to each other, then from Eqs. (1.2.25)–(1.2.27) we obtain

$$\left|S_{L,M}\right|^2 = F(\vec{\kappa}_{L,M})\Delta\kappa^2. \tag{1.2.28}$$

Note that the right-hand side of this equation is deterministic, whereas the left-hand side is the squared absolute value of a complex random parameter. This means that only the argument of this parameter is random, whereas the absolute value is the same for all terms of a stochastic ensemble.

In fact, the above mathematical manipulations indicate that the energy in an arbitrarily small but finite interval of spectral representation of a random process does not fluctuate, although the spectral amplitude of every harmonic and its absolute value are random numbers. Therefore, when replacing the continuous spectrum with the discrete one, if we want the harmonics of the discrete spectrum to replace the bands of the continuous spectrum and the intensity ratio of the harmonics to correspond to the energy ratio of the bands of the continuous spectrum, only the phase of the harmonics should be set in a random manner.

As for the distribution functions of phases (arguments) for the Fourier series coefficients, they can be chosen arbitrarily, because this choice has no effect on the form of the spectral density. However, it follows from the statistical independence of the real and imaginary parts that the arguments of the coefficients of the Fourier series are distributed uniformly in the interval $[0; 2\pi]$.

Let us consider some peculiarities connected with the fact that the process being considered is a real one. This fact is often used to obtain two phase screens as a result of a Fourier transform: one in the real part and another in the imaginary part. However, this approach is efficient only for a statistically uniform (for example, horizontal) path because only then can both screens have the same statistics. For vertical paths, the intensity of turbulence and its outer and inner scales, as well as the speed of phase screen transition (wind speed) vary along the path.

It is well known that the real component of a Fourier transform of a 1D real function is an even frequency function, whereas the imaginary component is an odd function; i.e.,

$$A(\omega) = A^{*}(-\omega)\,. \tag{1.2.29}$$

The Fourier transform of a 2D field has an analogous property:

$$A(\vec{\kappa}) = A^{*}(-\vec{\kappa})\,. \tag{1.2.30}$$

The symmetry property in Eq. (1.2.30) with respect to operation of conjugation for the discrete spectrum of the real function takes the form

$$\tilde{S}_{L,M} = \tilde{S}^{*}_{-L,\,-M}\,. \tag{1.2.31}$$

After calculation of the inner sum in Eq. (1.2.12) and row-by-row implementation of the Fourier transform, we obtain the matrix of counts:

$$\tilde{S}^x_{L,M} = \sum_{-N/2}^{N/2} S_{-L,-M} \exp\left(\frac{2\pi i}{N} L \times I\right), \tag{1.2.32}$$

where the property $\hat{S}^x_{L,M} = \tilde{S}^{x*}_{-L,-M}$. This allows us to calculate the inner sum for only non-negative values of M and thus obtain a twofold decrease in the number of mathematical operations.

Thus, the modification developed for the spectral sample method is the following sequence of operations:

1. Determine the first half of the rows of the 2D array of Fourier coefficients according to Eq. (1.2.28) for the modulo spectrum and the phase φ_{rand} uniformly distributed in the interval $[0; 2\pi]$:

$$\tilde{S}_{L,M} = F^{1/2}\left(\Delta\kappa\sqrt{L^2 + M^2}\right)\Delta\kappa \exp(i\varphi_{\mathrm{rand}}), \tag{1.2.33}$$

$$L = -N/2,\ N/2;\ M = 0,\ 1,\ \ldots,\ N/2.$$

2. Calculate the 1D Fourier transforms of these rows:

$$S^x_{L,M} = \sum_{-N/2}^{N/2} S_{L,M} \exp\left(\frac{2\pi i}{N} L \times I\right),\ M = 0,\ 1, \ldots,\ N/2. \tag{1.2.34}$$

3. Map of these transforms onto the second half of the rows with conjugation:

$$\tilde{S}^x_{I,M} = \tilde{S}^{x*}_{I,-M};\ M = -N/2, \ldots, -1. \tag{1.2.35}$$

4. Calculate the Fourier transform column by column:

$$S^x_{I,J} = \sum_{-N/2}^{N/2} \tilde{S}^x_{I,M} \exp\left(\frac{2\pi\iota}{N} M \times J\right),\ J = 0,\ 1, \ldots,\ N-1. \tag{1.2.36}$$

Within the framework of the "frozen turbulence" hypothesis, which can be formulated mathematically as follows:

$$S(\vec{\rho},t) = S(\vec{\rho} - \vec{V}t, 0), \tag{1.2.37}$$

this technique can be readily generalized to *dynamics problems*.

One of the properties of a Fourier transform is that the shift of the original function is equivalent to the addition of a linear component to the Fourier transform phase. Taking into account this property, Eq. (1.2.33) can be rewritten as

$$S_{L,M}(t) = F^{1/2}\left(\Delta\kappa\sqrt{L^2+M^2}\right)\Delta\kappa\exp(i\varphi_{\text{rand}})\exp\left[i\Delta\kappa t\left(V_x L + V_y M\right)\right],$$
$$L = -N/2,\ N/2;\ M = 0,\ 1,\ \ldots,\ N/2. \qquad (1.2.38)$$

Figure 1.2.1 shows an example of a series of phase screens representing turbulent distortions at the instants of time separated by $\Delta t = 10.2(\Delta\rho/V)$. The angle θ between the wind velocity and the abscissa was equal to 30 deg. As can be seen from this figure, turbulence moves along the wind velocity vector. Setting the direction so it does not coincide with the coordinate axis, we can obtain sufficiently long nonrecurring temporal realizations of turbulent distortions. This is possible because at every cycle along the abscissa with the duration $T_x = (N\ \Delta\rho)/V_x$, the phase screen shifts along the ordinate by $D_y = T_x V_y$. The period of exact recurrence is determined by the following condition:

$$nV_y T_x = mN\Delta\rho,$$

where n and m are even numbers. And, hence,

$$nV_y T_x = nV_y(N\Delta\rho)/V_x = mN\Delta\rho \Rightarrow nV_y = nV_x$$
$$\Rightarrow n = mV_x/V_y = \text{mctg}\theta. \qquad (1.2.39)$$

The technique was tested by comparing the integral square phase within a circle after subtraction of one or several aberrations ($N = 1, \ldots, 10$) described by the Zernike polynomials Z_l:

$$\Delta_N = \frac{1}{\pi R^2}\left\langle \int_{\rho<R} d^2\rho \left[S(\vec{\rho}) - \sum_{l=1}^{N} a_l Z_l(\vec{\rho})\right]^2 \right\rangle, \qquad (1.2.40)$$

with the corresponding parameters obtained theoretically in [42] and [43]. The calculated values of the integral variance of the approximation error obtained by averaging over 100 realizations are shown in Fig. 1.2.2 for a 128×128 computational grid and three different values of the ratio of the size L of the domain covered by the computational grid to the aperture diameter D. The calculated values are normalized to the corresponding analytically obtained values borrowed from Ref. [43]. The results are presented in Table 1.2.1.

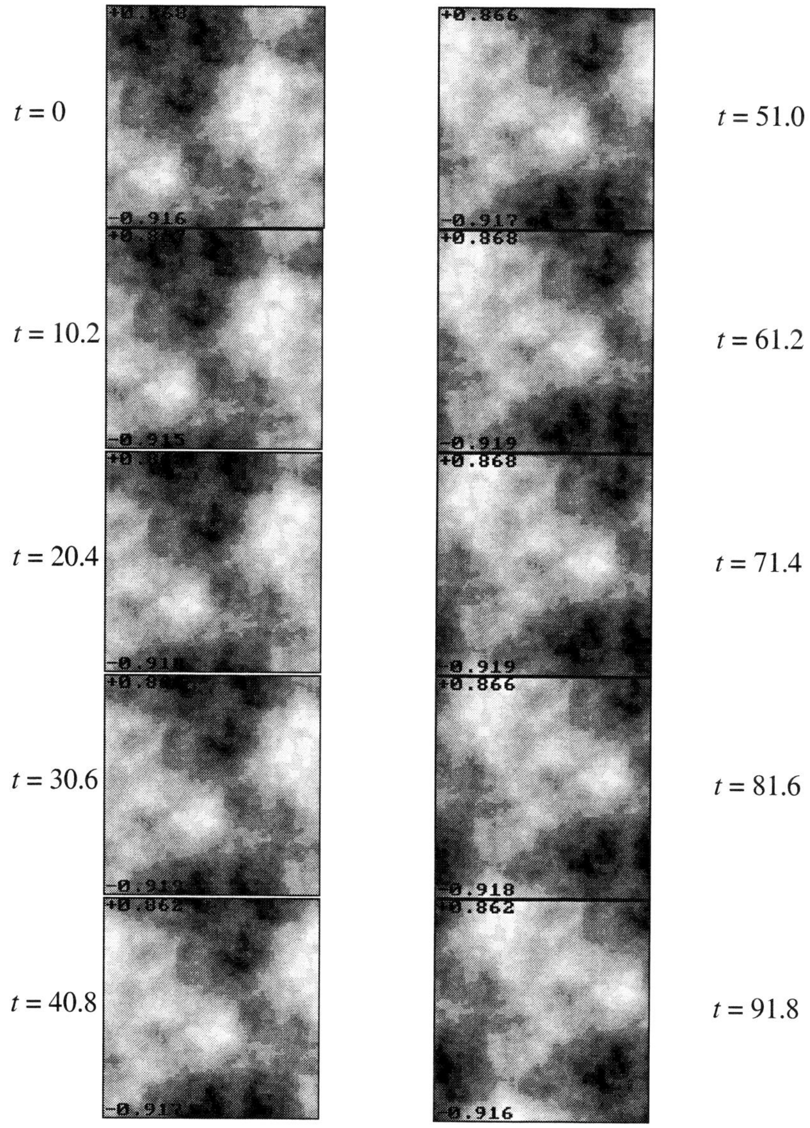

Figure 1.2.1. Series of phase screens representing turbulent distortions at time instants separated by the interval $\Delta t = 10.2\ (\Delta\rho/V)$.

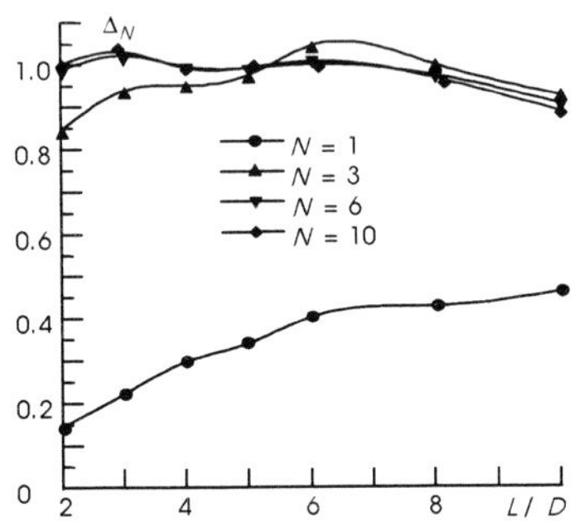

Figure 1.2.2. Calculated values of Δ_N from Eq. (1.2.40) normalized to the theoretical value for the power spectrum of turbulence.

Table 1.2.1. Analytically obtained integral variance of the error of approximating the turbulent wavefront by a finite number of Zernike polynomials.

Number of polynomials, N	$\Delta_N / (D / r_0)^{5/3}$
1	1.03
3	0.134
6	0.0648
8	0.0525

From the plots shown in Fig. 1.2.2 it can be seen that linear and quadratic aberrations are affected considerably by the loss of spatial scales larger than the size of the computational grid. This fact is also confirmed by calculation of the variance for the coefficients of wavefront expansion in terms of the Zernike polynomials for turbulent spectral density with the finite outer scale. The parameter Δ_1, which characterizes phase distortions minus the phase constant over the entire aperture (polynomial Z_1), is much less than the corresponding theoretical value. At the same time, for $L/D < 8$, the parameters Δ_6 and Δ_{10} differ from the theoretically obtained values by no more than 5%. For $L/D > 8$, the value of Δ_1 increases very slowly toward the theoretical limit, whereas parameters Δ_3, Δ_6, and Δ_{10} decrease at approximately the same speed, thus showing the poor approximation of high-order aberrations.

Thus, using the Fourier transform method to simulate a turbulent phase screen with the power spectrum, we obtain greatly underestimated values of tilt fluctuations (tilt corresponds to the first-order polynomials Z_2 and Z_3) and somewhat underestimated values for fluctuations of quadratic aberrations. This can be explained by the fact that the method is not suitable for simulation of large-scale random inhomogeneities and that the grid size L is an analog to the outer scale of turbulence.

1.3 Dynamic Simulation of the Large-Scale Part of Turbulent Aberrations of an Optical Phase

As mentioned earlier, the spectral sample method does not allow one to reproduce scales larger than the size of the computational grid. As a result, low spatial frequencies of the turbulent spectrum are lost, and this results in underestimation of atmospheric distortions, mainly those of random refraction and quadratic aberrations of the wavefront.

It is natural to try to compensate for underestimation of these aberrations. To do so, we need to know their statistics. Then we can sum two random phase screens: one generated by the Fourier transform method and including high and medium spatial frequencies, and another that includes aberrations induced by low spatial frequencies of the spectrum. This approach was formulated for the first time by Duncan and Collins in 1975 [7, 40]. In this section, we present a detailed justification and description of this approach, which is generalized here to the case of a time-dependent problem.

Since linear and quadratic aberrations coincide with the first terms of the Zernike series that is traditionally used in optics, it is convenient to apply the theory developed for this series; namely, the equations for statistical characteristics of Zernike coefficients that relate these statistical characteristics to the spectrum of a random process to be simulated.

Let us consider the representation of the wavefront as a Zernike series within a circular aperture with the radius R;

$$S(\vec{r}) = \sum_{l=0}^{\infty} a_l Z_l(\vec{r}) . \tag{1.3.1}$$

A Zernike polynomial Z_l [40-44] is a product of the radial $\left(R_n^m\right)$ and azimuth $\left(V_{m,l}\right)$ components:

$$Z_l = c_{mn} R_n^m(\vec{r}/R) V_{m,l}(\theta), \quad c_{mn} = 2^{\delta_{m0}/2}\sqrt{n+1}, \tag{1.3.2}$$

$$\begin{aligned} R_n^m(\vec{\rho}) &= \sum_{s=0}^{(n-m)/2} \frac{(-1)^s (n-m)!\rho^{n-2a}}{s![(n+m)/2-s]![(n-m)/2-s]!}, \\ V_{m,l} &= \cos(m\theta), \ l = 0,\ 2,\ 4,\ldots \\ V_{m,l} &= \sin(m\theta), \ l = 1,\ 3,\ 5,\ldots \end{aligned} \tag{1.3.3}$$

which are characterized by a radial power n and an azimuth frequency m so that the difference $n-m$ is an even number and $m \leq n$ (θ is the polar angle of the vector $\vec{r}$). In addition, Zernike polynomials are orthogonal within a circle; i.e.,

$$\iint_{r \leq R} Z_l(\vec{r}) Z_{l'}(\vec{r}) d^2 r = c_l \delta_{ll'}, \tag{1.3.4}$$

where $\delta_{ll'}$ is the Kronecker delta function.

The method of simulation for a random phase screen using the aberration representation consists of the following: first, we generate the vector of expansion coefficients and then this vector is substituted into the sum [Eq. (1.3.1)]. So, the problem is reduced to obtaining an ensemble of random Zernike coefficients. The simulated process is normal with a zero mean, therefore the expansion coefficients,

$$a_l = \frac{1}{\pi R^2} \iint_{r \leq R} S(\vec{r}) Z_l(\vec{r}) d^2 r, \tag{1.3.5}$$

are also Gaussian with a zero mean. To determine these coefficients, we have only to calculate the elements of the correlation matrix

$$C_{kl} = \langle a_k a_l \rangle = \frac{1}{\left(\pi R^2\right)^2} \iint_{r' \leq R} d^2 r' \iint_{r'' \leq R} d^2 r'' Z_k(\vec{r}') Z_k(\vec{r}'') \langle S(\vec{r}') S(\vec{r}'') \rangle . \tag{1.3.6}$$

Upon substitution of a random process $S(\vec{r})$ represented in the form of the Fourier–Stieltjes integral,

$$S(\vec{r}) = \iint_{-\infty}^{+\infty} \exp(i\vec{\kappa}\vec{r}) d^2 H(\vec{\kappa}) \quad , \tag{1.3.7}$$

alternation of the order of integration over spatial and frequency coordinates, and representation [44] of the Fourier transform of Zernike polynomials as

$$Q_l(\vec{\kappa}) = \frac{1}{\pi R^2} \iint_{r \leq R} d^2 r Z_l(\vec{r}) \exp(i\vec{\kappa}\vec{r}), \tag{1.3.8}$$

we obtain the following equation for elements of the correlation matrix C_{kl}:

$$C_{kl} = \int_{-\infty}^{+\infty} \int_{-\infty}^{+\infty} \int_{-\infty}^{+\infty} \int_{-\infty}^{+\infty} Q_k(\vec{\kappa}') Q_l^*(\vec{\kappa}'') \langle d^2 H(\vec{\kappa}') d^2 H^*(\vec{\kappa}'') \rangle . \tag{1.3.9}$$

Calculation of the integral [Eq. (1.3.8)] yields the following equation for representation of polynomials in the frequency region:

$$Q_l(\kappa,\Theta) = 2c_{nm}\, i^m (-1)^{(n-m)/2} \frac{J_{n+1}(\kappa R)}{\kappa R} \begin{Bmatrix} \cos(m\Theta) \\ \sin(m\Theta) \\ 1 \end{Bmatrix}. \qquad (1.3.10)$$

In this equation, Θ is the polar angle of the vector $\vec{\kappa}$.

With allowance made for delta correlation of the spectral components H,

$$\left\langle d^2H(\vec{\kappa}')\, d^2H^*(\vec{\kappa}'')\right\rangle = F(\vec{\kappa}')\, \delta(\vec{\kappa}'-\vec{\kappa}'')\, d^2\kappa' d^2\kappa'', \qquad (1.3.11)$$

upon integration of Eq. (1.3.9) over the angle of the vector of spatial frequency, we have

$$\begin{aligned} C_{l'l''} = {} & \delta_{m'm''} \times 8\pi[(n'+1)\times(n''+1)]^{1/2} \\ & \times(-1)^{(n'+n''-m'-m'')} \int_0^\infty F(\kappa) \frac{J_{n'+1}(\kappa R)\times J_{n''+1}(\kappa R)}{(\kappa R)^2} \kappa d\kappa, \end{aligned} \qquad (1.3.12)$$

where l' corresponds to n' and m', but l'' corresponds to n'' and m''. This equation determines the correlation of the coefficients of polynomials with the same azimuth factors. In other cases the correlation is zero.

For dynamic simulation, we also need to know the autocorrelation functions of the expansion coefficients

$$B_l(\tau) = \left\langle a_l(t) a_l^*(t+\tau)\right\rangle. \qquad (1.3.13)$$

On the assumption of the "frozen" field $S(\vec{\rho})$, i.e., at

$$S(\vec{r}, t+\tau) = S(\vec{r} - \vec{V}\tau, t), \qquad (1.3.14)$$

where $\vec{V} = (V, \vartheta)$ is the wind velocity (ϑ is the polar angle of the vector $\vec{V}$), we have

$$B_l(\tau) = \int_{-\infty}^{+\infty}\int_{-\infty}^{+\infty} |Q_l(\vec{\kappa})|^2 F(\vec{\kappa}) \exp\left(i\vec{\kappa}\vec{V}\tau\right) d^2\vec{\kappa}. \qquad (1.3.15)$$

Upon transition into the polar coordinate system and integration over the angular coordinate of the vector, we derive the following equation for $B_l(\tau)$:

$$B_l(\tau) = \langle a_l(t) a_l(t+\tau) \rangle = 8\pi(n+1)$$
$$\times \int_0^\infty \kappa d\kappa F(\kappa) \frac{J_{n+1}^2(\kappa R)}{(\kappa R)^2} \left[J_0(\kappa V \tau) + (-1)^{m+l} \cos(2m\vartheta) J_{2m}(\kappa V \tau)(1-\delta_{m0}) \right]. \quad (1.3.16)$$

For simulation of random temporal samples of the coefficients $a_l(t)$ according to the Fourier transform method, we need the analytical equation for the spectral density of fluctuations of the coefficients a_l, which is connected to the correlation function by the Wiener–Khintchine theorem:

$$W_l(\omega) = \frac{1}{\pi} \int_0^\infty d\tau B_l(\tau) \cos(\omega\tau), \quad (1.3.17)$$

where ω is the temporal frequency.

Substituting $B_l(\tau)$ from Eq. (1.3.17) into Eq. (1.3.16) and calculating the integral with respect to τ, we obtain the following equation:

$$W_l(\omega) = 8\pi(n+1) \times \int_{\omega/v}^\infty \kappa d\kappa F(\kappa) \frac{J_{n+1}^2(\kappa R)}{(\kappa R)^2} (\kappa^2 V^2 - \omega^2)^{-1/2}$$
$$\times \left\{ 1 - (1-\delta_{m0})(-1)^{m+l} \cos(2m\vartheta) \cos\left[\arcsin\left(\frac{\omega}{\kappa V} \right) \right] \right\}, \quad (1.3.18)$$

which can be used to generate random temporal realizations of the expansion coefficients. However, a difficulty arises that is due to correlation of the expansion coefficients in terms of the Zernike polynomials. The correlation matrix is nondiagonal, so we cannot generate coefficients as statistically independent random numbers. To solve this problem, we can use the expansion in the Karhunen–Loeve series [55], the correlation matrix of whose coefficients is diagonal by definition. However, the direct use of the Karhunen–Loeve expansion is problematic because the functions in this case have no explicit analytical form. It is possible to express the Zernike polynomials in terms of the Karhunen–Loeve functions through orthogonalization of the correlation matrix of the Zernike coefficients by a general similarity transformation. A detailed description of this procedure is given by Roddier in Ref. [45].

A more complicated problem arises in the case of dynamic simulation. Sometimes orthogonalization of the correlation matrix is impossible because polynomial coefficients with different angular components are correlated as well.

Note that the correlation function for the coefficients of two polynomials with the same radial n and angular m indices takes the form

$$B_{l'l''}(\tau)=8\pi(n+1)(-1)^{n-m}(-1)^{m}2^{\delta_{m0}-1}$$
$$\times\int_0^\infty \kappa d\kappa F(\kappa)\frac{J_{n+1}^2(\kappa R)}{(\kappa R)^2}J_{2m}(\kappa V\tau)\sin(2m\vartheta). \quad (1.3.19)$$

To break the correlation between these polynomials, it is sufficient to direct the abscissa of the coordinate system in which the polynomials are defined along the wind velocity vector. For this geometry,

$$\vartheta=\arctan(V_y/V_x)=0, \quad (1.3.20)$$

and the correlation of Eq. (1.3.19) is absent. This allows us to properly simulate the dynamics of the first- and second-order polynomial coefficients, i.e., tip and tilt, as well as those of defocusing and astigmatism (polynomials Z_2, ... Z_6).

In this case, small-scale aberrations can be simulated by the spectral sample method. The initial random process can be represented as a sum of the high- and low-frequency components S_{high} and S_{low}, whose spectra do not overlap:

$$S(\vec{r})=S_{\text{low}}(\vec{r})+S_{\text{high}}(\vec{r})=\iint_{|\vec{\kappa}|<\kappa_{\max}}\exp(i\vec{\kappa}\vec{r})d^2H(\vec{\kappa})+\iint_{|\vec{\kappa}|>\kappa_{\max}}\exp(i\vec{\kappa}\vec{r})d^2H(\vec{\kappa}). \quad (1.3.21)$$

Correlation of these terms is

$$\langle S_{\text{low}}(\vec{r}')S_{\text{high}}(\vec{r}'')\rangle=\iint_{|\kappa'|<\kappa_{\max}}\iint_{|\kappa''|>\kappa_{\max}}\exp[i(\vec{r}'\vec{\kappa}'-\vec{r}''\vec{\kappa}'')]\langle d^2H(\vec{\kappa}')d^2H^*(\vec{\kappa}'')\rangle$$
$$=\iint_{|\kappa'|<\kappa_{\max}}\iint_{|\kappa''|>\kappa_{\max}}\exp[i(\vec{r}'\vec{\kappa}'-\vec{r}''\vec{\kappa}'')]\delta(\vec{\kappa}'-\vec{\kappa}'')F(\kappa')\,d^2\kappa'd^2\kappa''=0 \quad (1.3.22)$$

owing to the properties of the delta function. With allowance for the fact that the processes are normal, we can consider them as statistically independent and simulate them separately, thus obtaining (after integration) a process with the required spectral density.

To implement this technique, we have developed a procedure for calculating the time spectra of the coefficient fluctuations of a Zernike expansion [Eq. (1.3.18)]. Figure 1.3.1 shows the spectra calculated for the polynomials of the first, second, and third radial orders. Temporal samples of the Zernike coefficients were generated according to the technique described in Sec. 1.2 for the 1D case. Figure 1.3.2 shows examples of realizations for the coefficients corresponding to tip and tilt, defocusing, and astigmatism. Figure 1.3.3 shows the dynamics of a phase screen obtained by summing the linear and quadratic aberrations with small-scale distortions generated by the spectral sample method described in Sec. 1.2.

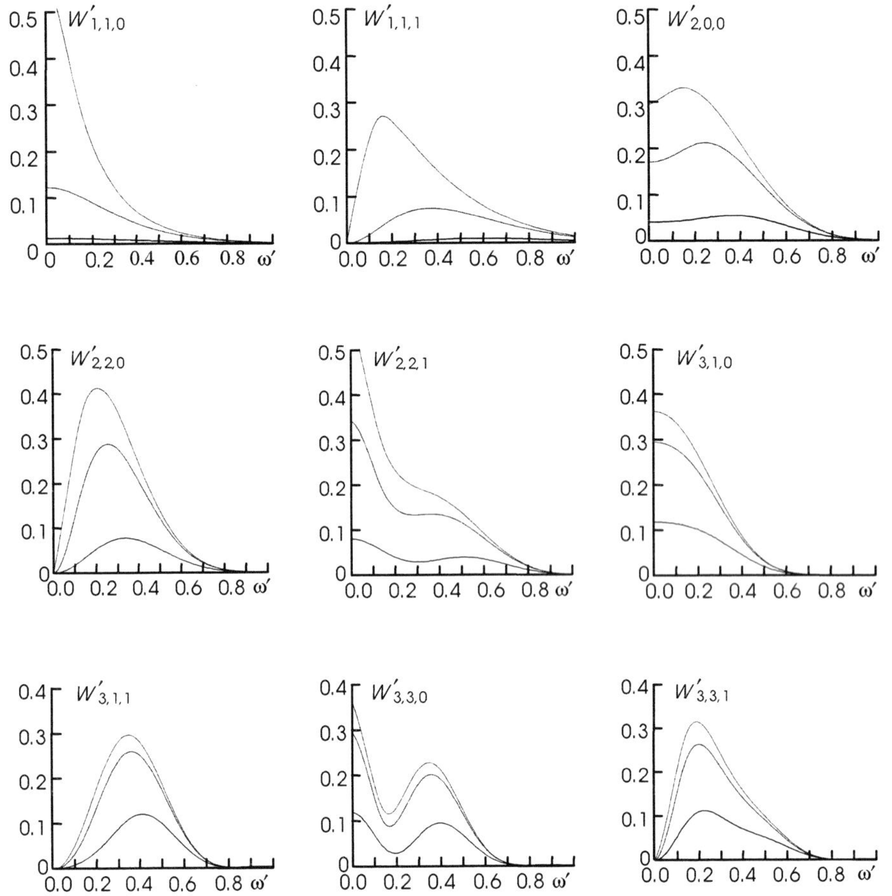

Figure 1.3.1. Spectra of fluctuations for the coefficients of Zernike polynomials of the first, second, and third radial orders at different values of the outer scale; $W' = W/\left[(1/2)\sigma_n^2\right]$ is the normalized spectrum; $\omega' = \omega/(3nV/R)$ is the normalized frequency; σ_n^2 is the variance of polynomial fluctuations for the infinite outer scale; n is the radial order of a polynomial; V is the wind velocity (directed along the 0x-axis); and R is the aperture radius. The three sets of subscripts to W' correspond to radial order, azimuth frequency, and parity parameter, respectively. In all figures the lower curves correspond to $L_0/D = 1$, the upper curves to $L_0/D = 10$, and the medium curves to $L_0/D = 3$; D is the aperture diameter.

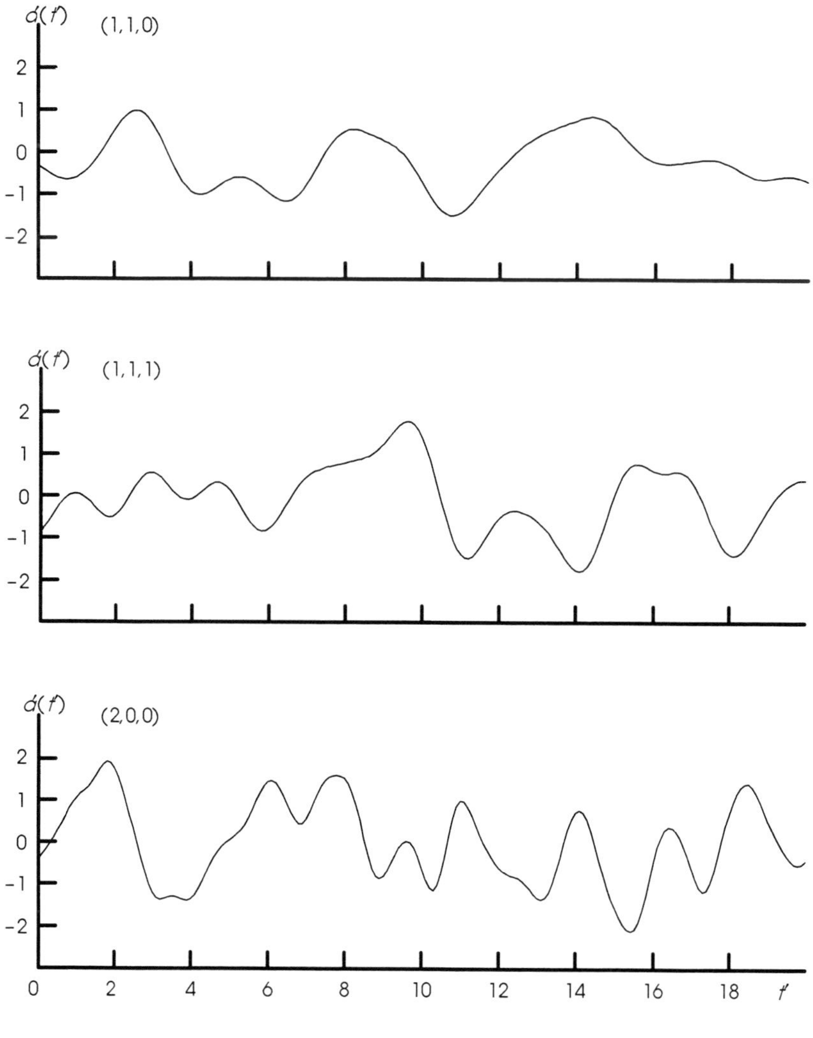

Figure 1.3.2. Random temporal samples of tilt and defocusing coefficients: $a'_n = a_n/\sigma_n$, $t' = t \times V/R$; σ_n is the standard deviation; $L_0/D = 10$. The numbers in parentheses above the curves correspond to radial order, azimuth frequency, and parity parameter.

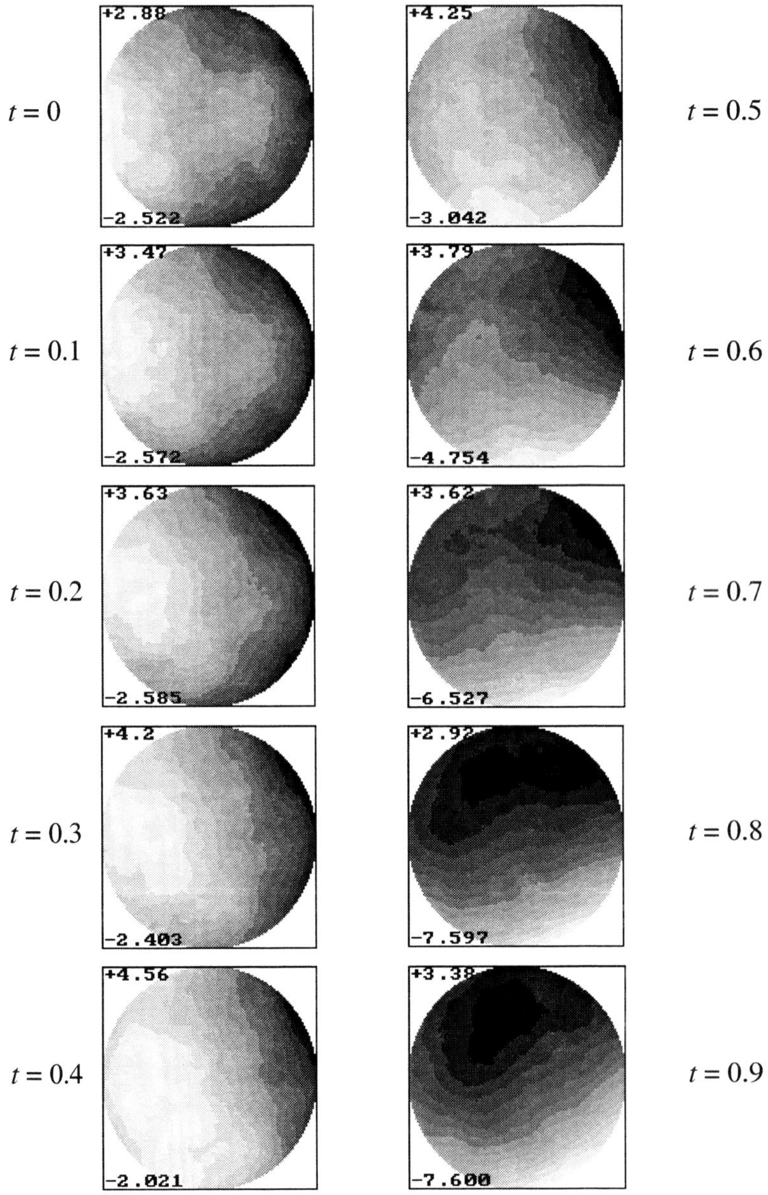

Figure 1.3.3. Dynamics of a phase screen obtained by summing linear and quadratic aberrations with small-scale distortions generated by the spectral sample method. The time is normalized to D/V; simulation was performed for the outer scale $L_0 = 50D$, $r_0 = D$ (r_0 is the Fried radius).

These calculations have been made for $L_0 = 50D = 100R$; i.e., for an outer scale two orders of magnitude larger than the aperture radius. A further increase in the outer scale leads to decreased accuracy during numerical calculation of the integral in Eq. (1.3.18). The precision of the simulation was verified by comparing the variance of the random realizations obtained with the variance calculated by Eq. (1.3.12). At $L_0 < 100R$ the difference did not exceed 5%, and at $L_0 > 100R$ it increased quickly because of the growing calculation error of the integral in Eq. (1.3.18).

1.4 Modification of the Numerical Model for Partially Coherent Beams

In the preceding sections we considered propagation of coherent beams. However, the radiation divergence for real laser sources is always greater than the diffraction limit. This is due to the processes that develop in the active medium of a laser, deformations of a laser cavity, the multimode structure of laser radiation, and so on.

Formally speaking, under the boundary conditions of the wave equation, we should allow for phase and amplitude fluctuations within the emitting aperture. Under time-dependent circumstances, this means the time dependence of the phase and amplitude [46, 47] is

$$E(\vec{\rho}, z=0) = \sqrt{I_0(\vec{\rho},t)}\exp[i\varphi(\vec{\rho},t)]. \tag{1.4.1}$$

In a stochastic representation, the phase and intensity become random functions of the transverse coordinates:

$$E(\vec{\rho}, z=0) = \sqrt{I_0(\vec{\rho})}\exp[i\varphi(\vec{\rho})]. \tag{1.4.2}$$

It is obvious that in both cases the statistical or dynamic description of the emitted radiation is unique for every type of laser source and even for a single laser. However, the temporal scales of fluctuations of the source intensity and phase are much less than that of the refractive index in the atmosphere. Thus, in our model we can omit the details of the space-time structure of partially coherent beams. Common practice in this case is to determine the boundary condition for the second-order coherence function:

$$\Gamma_2(\vec{\rho}_1,\vec{\rho}_2) = \left\langle E(\vec{\rho}_1)E^*(\vec{\rho}_2)\right\rangle. \tag{1.4.3}$$

If the statistics are uniform, we can introduce the coherence length in the following way:

$$\Gamma_2(\vec{\rho}_1,\vec{\rho}_2) = \Gamma_{20}(\vec{\rho}_1,\vec{\rho}_2)\exp(-|\vec{\rho}_1-\vec{\rho}_2|^2/\rho_c^2), \tag{1.4.4}$$

where $\Gamma_{20}(\vec{\rho}_1, \vec{\rho}_2)$ corresponds to the completely coherent radiation $(\rho_c = \infty)$:

$$\Gamma_{20}(\vec{\rho}_1, \vec{\rho}_2) = U_0(\vec{\rho}_1) U_2^*(\vec{\rho}_2), \tag{1.4.5}$$

and where $U_0(\vec{\rho})$ is a regular component of the field.

In the mathematical model employed, we have to solve the wave equation describing the field. In the framework of this model, we need a method that allows propagation of partially coherent radiation. Note that partial coherence is equivalent to additional angular divergence of a beam. Equations for the effective size a of a partially coherent Gaussian beam propagating in a turbulent atmosphere can be written as [48]

$$a^2 = a_g^2 + a_d^2 + a_c^2 + a_t^2. \tag{1.4.6}$$

Here,

$$a_g = a_0(1 - z/F) \tag{1.4.7}$$

is the size of the beam focused at a distance F in the cross section z, and a_0 is the initial effective radius of the Gaussian beam;

$$a_d = z/ka_0 = z \times \frac{\lambda}{D}, \quad D = 2\pi a_0 \tag{1.4.8}$$

is the cross section of a cone with a base equal to the diffraction divergence $1/ka_0$;

$$a_c = z/k\rho_c \tag{1.4.9}$$

is the cross section of a cone with a base equal to the divergence of the partially coherent beam, and

$$a_t = z/k\rho_0 \tag{1.4.10}$$

is the cross section of a cone with a base inversely proportional to the coherence length of the turbulent phase screen ρ_0.

Note that the last equation addresses the problem of beam propagation behind a random phase screen. For a beam propagating in a randomly inhomogeneous medium, turbulent divergence is a function of longitudinal coordinate z. So, to take into account the different factors that influence beam propagation, we should sum the corresponding cross sections. Naturally, the effect of the different factors is assumed to be independent.

In the following discussion we propose a modification of the numerical model for propagation of a coherent beam to allow for the initial divergence of the beam. In the case of a random phase screen, the result obtained coincides absolutely with the analytical equation written above.

Consider now the proposed modification. As described earlier, the wave equation is solved for a coherent beam. However, the intensity in the cross sections is calculated as a convolution of the intensity of coherent radiation, with function W describing beam broadening that is due to partial coherence of the source:

$$I(\vec{\rho},z) = |E(\vec{\rho},z)|^2 \otimes W(\vec{\rho},z), \qquad (1.4.11)$$

where $\otimes$ refers to convolution. Assuming that W is a Gaussian function with the width $a_c = z/k\rho_c$ and that

$$W(\vec{\rho},z) = \exp\left(-\frac{\rho^2}{a_c^2}\right) = \exp\left(-\frac{\rho^2}{z^2}k^2\rho_c^2\right), \qquad (1.4.12)$$

and solving the problem of diffraction of a partially coherent beam in a vacuum, we obtain results that are exactly the same as those obtained with the use of the analytical equation. Indeed, convolution of two Gaussian functions with widths a_1 and a_2 is a Gaussian function with the square width equal to the sum of the square widths of these two functions:

$$\exp\left(-\rho^2/a_1^2\right) \otimes \exp\left(-\rho^2/a_2^2\right) = \text{const}\exp\left[-\rho^2/\left(a_1^2+a_2^2\right)\right]. \qquad (1.4.13)$$

This equation can be readily proved by calculating convolution through a Fourier transform.

Therefore, Eq. (1.4.6) can be considered as a sequential convolution (in the beam cross section) of the initial Gaussian intensity distribution obtained by the ray-optics approximation, with functions that describe diffraction, partial coherence, or turbulent divergence.

Another interpretation of the problem is possible as well. Let us assume that partial coherence of the beam is due to high-frequency fluctuations of the beam direction, and the angle β between the beam direction and the axis of the optical system is a random value. Then, the distribution of the mean intensity in the beam cross section at a distance z from the source is

$$\langle I(\vec{\rho},z)\rangle = \iint I\left(\vec{\rho}-\vec{\beta}z,z\right) f(\vec{\beta})d^2\beta = I(\vec{\rho},z) \otimes f(\vec{\rho}/z). \qquad (1.4.14)$$

Here $f(\vec{\beta})$ is a two-dimensional function of the probability density of the beam direction and $I(\vec{\rho},z)$ is the intensity distribution in the cross section of the unperturbed beam. The second-order coherence function is

$$\Gamma_2(\vec{\rho}_1,\vec{\rho}_2)=\Gamma_{20}(\vec{\rho}_1,\vec{\rho}_2)\langle\exp[ik\beta_x(x_1-x_2)]\rangle \times\left\langle\exp\left[ik\beta_y(y_1-y_2)\right]\right\rangle=\Gamma_{20}(\vec{\rho}_1,\vec{\rho}_2)\exp\left[-k^2\sigma_\beta^2(\vec{\rho}_1-\vec{\rho}_2)^2/2\right] \tag{1.4.15}$$

on the assumption that the beam direction is a two-dimensional normally distributed random function with the variance σ_β^2 in every realization, and that the correlation between the components of the vector $\vec{\beta}$ is absent. It follows from the previous equation that the relationship between the coherence length and the mean-square value of $\vec{\beta}$ is

$$\rho_c=\frac{2}{k\sigma_\beta}, \text{ or } \sigma_\beta=\frac{2}{k\rho_c}. \tag{1.4.16}$$

Thus, by considering the case in which partial coherence is caused by beam jitter, we have also obtained the equation in which the mean intensity of the partially coherent beam is determined by convolution of the diffraction-limited solution of the wave equation with the function describing additional beam broadening.

Let us also note that when deriving the convolution equation we did not impose any restrictions on the intensity distribution in the beam cross section or on the probability density function of random wandering of the beam direction. This equation is valid for non-Gaussian beams as well. Moreover, we did not make any assumptions concerning the phase of the regular field component, so the equation is also suitable for describing beam propagation behind a random or nonlinear phase screen.

Overall, the propagation of a high-power partially coherent laser beam is described by the following set of equations:

$$\begin{cases} 2ik\dfrac{\partial E}{\partial z}=\left(\dfrac{\partial^2}{\partial x^2}+\dfrac{\partial^2}{\partial y^2}+2k^2\delta n\right)E, \\ M(\delta n)=\alpha\left|E\times E^*\right|\otimes W(\vec{\rho},\sigma_\beta,z), \\ W(\vec{\rho},\sigma_\beta,z)=\exp\left(\dfrac{-\dfrac{\rho^2}{\sigma_\beta^2 z^2}}{C_{\text{norm}}}\right), \\ E(\vec{\rho},z=0)=U_0(\vec{\rho}). \end{cases} \tag{1.4.17}$$

As defined earlier, M is the operator that describes variations of the refractive index that are due to absorption of the beam energy and U_0 is the regular field component (i.e., $\rho_c=\infty$). The parameter σ_β^2 is the angular divergence that is due to partial coherence (this divergence is not necessarily caused by random fluctuations of the beam direction); C_{norm} is the normalization constant.

According to the splitting algorithm [49, 50], convolutions are calculated only in the cross sections corresponding to the coordinates of the nonlinear phase screen and in the plane of the receiving aperture.

In order to properly estimate thermal blooming of a beam in a thick layer of a randomly inhomogeneous medium, further investigation of the proposed model is needed. It is possible that the form of the function W depends on the mechanism limiting the coherence length. However, the general features of our approach remain if the source field is averaged over fluctuations more quickly than variations of the refractive index.

1.5 Lens Transformation of Coordinates in an Inhomogeneous Wave Equation

In some circumstances, a transformation of coordinates is needed for optical beams propagating in random media. There is one interesting approach that uses lens transformation. Lens transformation of coordinates, known as Talanov transformation, has been used by several authors [2, 39, 51-54] to transform the problem of propagation of a beam focused by an aperture with one Fresnel number into the same problem but with an aperture with a different Fresnel number. During numerical solution of the problem of paraxial beam propagation, this transformation allows the computational expense to be decreased markedly, particularly in the case of large Fresnel numbers (small wave parameters). Here we give the rigorous deduction of the equation for this transformation, which allows construction of this transformation as applied to complex cases of interaction of a high-power beam with a medium. This is demonstrated in this section using nonstationary thermal blooming under conditions of forced convection with allowance made for thermal conductivity of the medium.

Consider first the case of *free diffraction*. Propagation of the complex amplitude $U(\vec{\rho}, z)$ in the direction of the $0z$-axis is described by the homogeneous wave equation:

$$2ik\frac{\partial U}{\partial z} = \nabla_{\perp}^{2} U . \tag{1.5.1}$$

Solution of this wave equation by the Fourier transform method or Green's functions yields the well-known convolution integral:

$$U(\vec{\rho}, z) = \frac{i}{\lambda z}\iint d^{2}\rho' U_{0}(\vec{\rho}')\exp\left[-\frac{ik}{2z}(\vec{\rho} - \vec{\rho}')^{2}\right] . \tag{1.5.2}$$

Consider two thin collecting lenses L_1 and L_2 with the focal lengths f_1 and f_2. Let the wave with the complex amplitude $U_0(\vec{\rho})$ be incident on the entrance pupil of each lens. Then, for the distance z_1 behind the plane of the lens L_1 we have the following equation for the field U_1:

$$U_1(\vec{\rho}, z_1) = \frac{i}{\lambda z_1} \exp\left(-\frac{ik}{2z_1}\rho^2\right) \iint d^2\rho' U_0(\vec{\rho}') \exp\left(i\frac{k}{z_1}\vec{\rho}\vec{\rho}'\right) \exp\left[\frac{ik}{2}\rho'^2\left(\frac{1}{f_1} - \frac{1}{z_1}\right)\right]. \tag{1.5.3}$$

Similarly, at the distance z_2 behind the plane of the second lens,

$$U_2(\vec{\rho}, z_2) = \frac{i}{\lambda z_2} \exp\left(-\frac{ik}{2z_2}\rho^2\right) \iint d^2\rho' U_0(\vec{\rho}') \exp\left(i\frac{k}{z_2}\vec{\rho}\vec{\rho}'\right) \exp\left[\frac{ik}{2}\rho'^2\left(\frac{1}{f_2} - \frac{1}{z_2}\right)\right]. \tag{1.5.4}$$

In the particular case of $z_1 = f_1$ and $z_2 = f_2$, it follows from Eqs. (1.5.3)–(1.5.4) that the fields in the focal planes are related to each other by the following equation:

$$U_2(\vec{\rho}, f_2) = \frac{f_1}{f_2} U_1\left(\vec{\rho}\frac{f_1}{f_2}, f_1\right) \exp\left[-\frac{i}{2}\rho^2 k\left(\frac{1}{f_1} - \frac{1}{f_2}\right)\right], \tag{1.5.5}$$

and the corresponding intensity distributions differ by a factor and a scale

$$I_2(\vec{\rho}, f_2) = (f_1/f_2)^2 I_1(\rho f_1/f_2). \tag{1.5.6}$$

Generally, one field can be expressed through another if the following condition is fulfilled:

$$1/f_1 - 1/z_1 = 1/f_2 - 1/z_2 . \tag{1.5.7}$$

Upon designation of $\delta = 1/f_2 - 1/f_1$, we obtain

$$z_1 = z_2/(1 - \delta z_2), \tag{1.5.8}$$

$$z_2 = z_1/(1 + \delta z_1). \tag{1.5.9}$$

The fields in these cross sections are related by the equation

$$U_2(\vec{\rho}, z_2) = \frac{1}{(1 - \delta z_2)} U_1\left[\frac{\vec{\rho}}{(1 - \delta z_2)}, z_1\right] \exp\left[\frac{ik}{2}\rho^2 \frac{\delta}{(1 - \delta z_2)}\right]. \tag{1.5.10}$$

Substituting Eq. (1.5.8) into this equation and omitting the subscript of z, we have

$$U_2(\vec{\rho}, z) = \frac{1}{(1 - \delta z)} U_1\left[\frac{\vec{\rho}}{(1 - \delta z)}, \frac{z}{(1 - \delta z)}\right] \exp\left[\frac{ik}{2}\rho^2 \frac{\delta}{(1 - \delta z)}\right] \tag{1.5.11}$$

and similarly

$$U_1(\vec{\rho},z)=\frac{1}{(1+\delta z)}U_2\left[\frac{\vec{\rho}}{(1+\delta z)},\ \frac{z}{(1+\delta z)}\right]\exp\left[\frac{ik}{2}\rho^2\frac{-\delta}{(1+\delta z)}\right]. \quad (1.5.12)$$

Thus, the fields, which differ in the plane $z=0$ by the square phase factor $\exp(1/2\,ik\delta\rho^2)$, are related by Eqs. (1.5.11) and (1.5.12) *at diffraction in a free space or an optically homogeneous medium.* These equations allow one field to be expressed through another in the cross sections meeting the condition of Eq. (1.5.7).

In an optically inhomogeneous medium, in which the refractive index is a function of coordinates, the propagation of the complex amplitude is described by the inhomogeneous parabolic equation:

$$2ik\frac{\partial U}{\partial z}=\nabla_\perp^2 U+2k^2\tilde{n}(\vec{\rho},z)U\,. \quad (1.5.13)$$

Let the boundary conditions be given in the plane $z=0$ as

$$U_1(\vec{\rho},0)=U_0(\vec{\rho}), \quad (1.5.14)$$

$$U_2(\vec{\rho},0)=U_0(\vec{\rho})\exp\left[\frac{1}{2}ik\delta\rho^2\right]. \quad (1.5.15)$$

As has been shown earlier, while propagating in a homogeneous medium, these fields are related by Eqs. (1.5.11) and (1.5.12). Let the field U_1 propagate in a medium with the refractive index $n_1(\vec{\rho},z)$; then Eq. (1.5.13) can be written as

$$2ik(\partial U_1/\partial z)=\nabla_\perp^2 U_1+2k^2\tilde{n}_1(\vec{\rho},z)U_1\,. \quad (1.5.16)$$

Our purpose is to find a distribution of the refractive index $n_2(\vec{\rho},z)$ such that with propagation of the wave U_2 meeting the equation

$$2ik(\partial U_2/\partial z)=\nabla_\perp^2 U_2+2k^2\tilde{n}_2(\vec{\rho},z)U_2; \quad (1.5.17)$$

Eqs. (1.5.11) and (1.5.12) remain valid.

For convenience, rewrite the last two equations as

$$2k^2\tilde{n}_2(\vec{r})U_2(\vec{r})=2ikU'_{2z}(\vec{r})-U''_{2xx}(\vec{r})-U''_{2yy}(\vec{r}), \quad (1.5.18)$$

$$2k^2\tilde{n}_1(\vec{r})U_1(\vec{r})=2ikU'_{1z}(\vec{r})-U''_{1xx}(\vec{r})-U''_{1yy}(\vec{r}), \quad (1.5.19)$$

using

$$U'_{1z}(\vec{r})=\partial U_1/\partial z, \qquad U'_{2z}(\vec{r})=\partial U_2/\partial z, U''_{1xx}(\vec{r})=\partial^2 U_1/\partial x^2,$$
$$U''_{2xx}(\vec{r})=\partial^2 U_2/\partial x^2, \quad U''_{1yy}(\vec{r})=\partial^2 U_1/\partial y^2, \quad U''_{2yy}(\vec{r})=\partial^2 U_2/\partial y^2.$$

Without loss of generality, the latter can be written in the transformed coordinate system as

$$\begin{aligned} &2k^2\tilde{n}_1[\vec{r}/(1-\delta z)]U_1[\vec{r}/(1-\delta z)] \\ &= 2ikU'_{1z}[\vec{r}/(1-\delta z)] - U''_{1xx}[\vec{r}/(1-\delta z)] - U''_{1yy}[\vec{r}/(1-\delta z)]. \end{aligned} \tag{1.5.20}$$

Dividing Eq. (1.5.18) by Eq. (1.5.20), we get

$$\begin{aligned} \frac{n_2(\vec{r})}{n_1[\vec{r}/(1-\delta z)]} &= \frac{U_1[\vec{r}/(1-\delta z)]}{U_2(\vec{r})} \\ &\times \frac{2ikU'_{2z}(\vec{r}) - U''_{2xx}(\vec{r}) - U''_{2yy}(\vec{r})}{2ikU'_{1z}[\vec{r}/(1-\delta z)] - U''_{1xx}[\vec{r}/(1-\delta z)] - U''_{1yy}[\vec{r}/(1-\delta z)]}. \end{aligned} \tag{1.5.21}$$

The first fraction on the right-hand side of Eq. (1.5.21) follows from Eq. (1.5.11):

$$\frac{U_1[\vec{r}/(1-\delta z)]}{U_2(\vec{r})} = (1-\delta z)\exp\left[-\frac{1}{2}ik\rho^2\frac{\delta}{(1-\delta z)}\right]. \tag{1.5.22}$$

To calculate the second fraction, differentiate Eq. (1.5.11) with respect to ∂z, ∂x^2, and ∂y^2. Combining the results of differentiation, we can then derive the following equation:

$$\begin{aligned} &2ikU'_{2z}(\vec{r}) - U''_{2xx}(\vec{r}) - U''_{2yy}(\vec{r}) = \frac{1}{(1-\delta z)^3}\exp\left[\frac{1}{2}k\rho^2\Big/(1-\delta z)\right] \\ &\times\left\{2ikU'_{1z}[\vec{r}/(1-\delta z)] - U''_{1xx}[\vec{r}/(1-\delta z)] - U''_{1yy}[\vec{r}/(1-\delta z)]\right\}, \end{aligned} \tag{1.5.23}$$

from which it is seen that the second fraction on the right-hand side of Eq. (1.5.21) is equal to

$$\frac{\exp\left[\frac{1}{2}k\rho^2\Big/(1-\delta z)\right]}{(1-\delta z)^3}. \tag{1.5.24}$$

Substituting Eqs. (1.5.22) and (1.5.24) into Eq. (1.5.21), we find the relation sought:

$$n_2(\vec{r}) = \frac{1}{(1-\delta z)^2} \times n_1[\vec{r}/(1-\delta z)]. \quad (1.5.25)$$

Let us use this relation in a problem of thermal blooming. In an isobaric approximation, the temperature field T is described by the forced heat transfer equation:

$$\frac{\partial T}{\partial t} + \vec{V}(z)\nabla_\perp T - \chi(z)\nabla_\perp^2 T = \frac{\alpha}{\rho C_P} I(\vec{r}), \quad (1.5.26)$$

where α is the absorption coefficient, ρ_0 is the medium density, C_P is specific heat at constant pressure, $\vec{V}(z)$ is the wind velocity profile, and χ is heat conductivity.

Assuming the refractive index to be a linear function of temperature, i.e., at $n - n_0 \approx n'_T (T - T_0)$, we obtain the following differential equation for it:

$$n'_t + \vec{V}(z)\nabla_\perp n - \chi(z)\nabla_\perp^2 n = q(z) \times I(\vec{r}), \quad (1.5.27)$$

where $q(z) = \dfrac{\alpha(z) n'_t}{\rho_0 C_P}$, $n'_t = \partial n / \partial t$.

Then consider the propagation of two high-power optical beams with the boundary conditions of Eqs. (1.5.14) and (1.5.15). We have the parameters of the path at which thermal blooming of the first beam occurs: $\vec{V}_1(z)$, $\chi_1(z)$, and $\alpha_1(z)$. The problem is to determine the parameters of the second path (with parameters $\vec{V}_2(z)$, $\chi_1(z)$, and $\alpha_1(z)$ so that condition (1.5.25) is fulfilled and the complex amplitudes of the beams are related by Eqs. (1.5.11) and (1.5.12).

Divide Eq. (1.5.27) written for $n_1[\vec{r}/(1-\delta z)]$ by Eq. (1.5.25) for $n_2(\vec{r})$. With allowance made for the intensity cross section ratio being the consequence of Eq. (1.5.6):

$$I_1[\vec{r}/(1-\delta z)]/I_2(\vec{r}) = (1-\delta z)^2, \quad (1.5.28)$$

and the results of differentiation of Eq. (1.5.25):

$$\begin{aligned} n'_{1t}[\vec{r}/(1-\delta z)] &= (1-\delta z)^2 \times n'_{2t}(\vec{r}) \\ \vec{\nabla}_\perp n_1[\vec{r}/(1-\delta z)] &= (1-\delta z)^3 \times \vec{\nabla}_\perp n_1(\vec{r}) \\ \nabla_\perp^2 n_1[\vec{r}/(1-\delta z)] &= (1-\delta z)^4 \times \nabla_\perp^2 n_1(\vec{r}), \end{aligned} \quad (1.5.29)$$

we obtain the following equation:

$$\frac{n'_{2t} + (1-\delta z)\times \vec{V}_1[z/(1-\delta z)]\nabla_\perp n_2 - (1-\delta z)^2 \chi_1[z/(1-\delta z)]\nabla_\perp^2 n_2}{n'_{2t} + \vec{V}_2(z)\nabla_\perp n_2 - \chi_2(z)\nabla_\perp^2 n_2} = \frac{\alpha_1[z/(1-\delta z)]}{\alpha_2(z)}. \tag{1.5.30}$$

It can be seen that the equality is fulfilled in the following relation between the path parameters:

$$\alpha_2(z) = \alpha_1[z/(1-\delta z)]; \tag{1.5.31}$$

$$\vec{V}_2(z) = (1-\delta z)\times \vec{V}_1[z/(1-\delta z)]; \tag{1.5.32}$$

$$\chi_2(z) = (1-\delta z)^2 \times \chi_1[z/(1-\delta z)]. \tag{1.5.33}$$

Equations (1.5.31)–(1.5.33) give the profiles of absorption, wind velocity, and thermal conductivity at which the solution of the problem of thermal blooming for a beam with a focal length f_2 at every instant can be found in the solution of the initial problem for a beam focused at a distance f_1 (by the lens transformation equations).

In this chapter, we have considered various aspects of numerical simulation of optical wave propagation through the atmosphere. In the following chapters these results will be used as a basis for studying the efficiency of various schemes and algorithms for minimizing atmospheric distortions in beaming and imaging.

References

1. R. Buckley, "Diffraction by a random phase-changing screen: A numerical experiment," *J. Atm. Terrestrial Phys.*, **37**, No.12, pp. 1431–1446, 1975.
2. J.A. Fleck, J.R. Morris, M.D. Feit, "Time-dependent propagation of a high-energy laser beam through the atmosphere," *Appl. Phys.*, **10**, No. 1, pp. 129–139, 1976.
3. V.P. Kandidov, V.I. Ledenev, "Application of the method of statistical testing for investigation of wave beam propagation in statistically random media," *I.V.U.Z Ser. Radiofiz.*, **24**, No. 4, pp. 434–442, 1981.
4. P.A. Konyaev, "Modification of the splitting method for numerical solution of quasi-optical problems," in *Abstracts of Reports at the VI Conference on Laser Beam Propagation in the Atmosphere*, pp. 195–198, 1981.
5. J.M. Martin, S.M. Flatte, "Intensity images and statistics from numerical simulation of wave propagation in 3-D random media," *Appl. Opt.*, **27**, No.11, pp. 2111–2126, 1988.
6. V.V. Bykov, *Numerical Simulation in Statistical Radiotechnics*, Sov. Radio, Moscow, 1971.

7. S.A. Collins [Jr.] and D.D. Duncan, "Simulation atmospherically degraded beams. II. Polynomial approach," *J. Opt. Soc. Am.*, **65**, No.10, p.1218, 1975.
8. S.A. Collins Jr., D.D. Duncan, "Simulation of atmospherically degraded beams. I. Extended Fourier approach," *J. Opt. Soc. Am.*, **65**, No.10, 1975.
9. I.E. Tel'pukhovski, S.S. Chesnokov, "Model presentation of atmospheric distortions for numerical analysis of statistical characteristics of optical beams," *Atm. Opt.*, **4**, pp.1294–1297, 1991.
10. B.V. Fortes, V.P. Lukin, "Modeling of an image observed through a turbulent atmosphere," *Proc. SPIE Vol.* **1688**, pp. 477–488, 1992.
11. J. Rickling, *Modern Optics*, 1995.
12. G.W. Carhart, M.A. Vorontsov, "Adaptive aberration correction based on opto-electronic Zernike wavefront sensor and the decoupled stochastic parallel gradient descent control technique," *Proc. SPIE Vol.* **4493**, 2001.
13. M. Roggermann, B.M. Welsh, *Imaging Through Turbulence*, CRC Press, Boca Raton, FL, 1996.
14. J.M. Martin, M. Flatte, "Simulation of point-source scintillation through three-dimensional random media," *J. Opt. Soc. Am. A* **7**, pp. 838–847, 1990.
15. V.A. Vysloukh, V.P. Kandidov, S.S. Chesnokov, S.A. Shlenov, "Adaptive focusing of intensive light beams propagating in non-regular media," *Fizika*, **228**, pp. 30–41, 1985.
16. V.P. Lukin, P.A. Konyaev, "Thermal distortions of focused laser beams in the atmosphere," *Appl. Opt.*, **24**, pp. 415–421, 1985.
17. V.P. Lukin, B.V. Fortes, "Phase distortions of an optical beam due to its self-action under conditions of gravitational convection," *Atm. Opt.*, **3**, pp. 1182–1185, 1990.
18. V.P .Lukin, F.Yu. Kanev, P.A. Konyaev, B.V. Fortes, "Numerical model of an atmospheric adaptive optical system. Parts I, II, III," *Atm. Oceanic Opt.*, **8**, pp. 210–222, 1995.
19. V.I . Tatarskii, *Wave Propagation in the Turbulent Atmosphere*, Nauka, Moscow, 1967.
20. A. Ishimaru, *Wave Propagation and Scattering in Random Media,* Academic Press, New York, 1977.
21. J.W. Strohben, ed., *Laser Beam Propagation in the Atmosphere*, Springer-Verlag, Berlin-Heidelberg-New York, 1978.
22. J.L. Walsh, P.B. Ulrich, in *Laser Beam Propagation in the Atmosphere*, edited by J.W. Strohben, Springer-Verlag, Berlin-Heidelberg-New York, 1978.
23. G.I. Marchuk, *Methods of Numerical Mathematics*, Nauka, Moscow, 1980.
24. P.A. Konyaev, "Numerical solution of diffraction problems on a random phase screen," in *Abstracts of Reports at the V Conference on Laser Beam Propagation in the Atmosphere*, pp. 120–122, 1979.
25. S.M. Rytov, *Introduction to Statistical Radiophysics*, Nauka, Moscow, 1966.
26. V.V. Vorob'iev, *Thermal Blooming of Laser Beams in the Atmosphere: Theory and Model Experiment*, Nauka, Moscow, 1987.

27. B.M. Berkovskii, E.F. Nogotov, *Numerical Experiment in the Problems of Convections*, University Publishing House, Minsk, 1988.
28. A.A. Samarskii, *Introduction to the Theory of Finite-Difference Schemes*, Nauka, Moscow, 1971.
29. V.P. Lukin, "Adaptive formation of beams and images in the turbulent atmosphere," *Atm. Oceanic Opt.*, **8**, pp. 152–173, 1995.
30. V.E. Zuev, P.A. Konyaev, V.P. Lukin, "Minimization of atmospheric distortions of optical waves by methods of adaptive optics," *Fizika*, **28**, pp. 6–29, 1985.
31. V.P. Kandidov, M.P. Tamarov, S.A. Shlyonov, "Spatial statistics of laser beams in small-scale turbulence. Stochastical simulation," *Atm. Oceanic Opt.*, **9**, pp. 1443–1449, 1996.
32. B.V. Fortes, V.P. Lukin, *Adaptive Formation of Beams and Images in the Atmosphere*, Publishing House of the Siberian Branch of the Russian Academy of Sciences, Novosibirsk, 1999.
33. A.N. Kolmogorov, "Local structure of turbulence in liquid under very large Rayleigh numbers," *Dok. Akad. Nauk SSR*, **30**, pp. 299–303, 1941.
34. A.M. Obukhov, "Energy distribution in spectra of a turbulent stream," *Dok. Akad. Nauk SSR*, **32**, pp. 22–24, 1943.
35. V.P. Lukin, *Atmospheric Adaptive Optics*, Nauka, Novosibirsk, 1986.
36. A.M.Monin, A.M. Yaglom, *Statistical Hydrodynamics. Part II*, Nauka, Moscow, 1967.
37. L.I. Mirkin, M.A. Rabinovich, L.P. Jaroslavski, "Method of generation of correlated Gaussian pseudo-random numbers on a computer," *J. Com. Math. and Math. Physics*, **12,** pp. 1353–1357, 1972.
38. R.B. Shack, B.C. Peri, "Production and use of a lenticular Hartmann screen," *J. Opt. Soc. Am.*, **6**, p. 1586, 1971.
39. E.A. Sziclas, A.E. Siegman, "Diffraction calculation using fast Fourier transform methods," *Proc. IEEE*, **62**, No. 3, pp. 410–412, 1974.
40. J.A. Fleck, J.R. Morris, "Equivalent thin lens model for thermal blooming compensation," *Appl. Opt.*, **17**, pp. 2575–2579, 1978.
41. M. Born and E. Wolf, *Principles of Optics*, Pergamon, New York, 1959.
42. R.J. Noll, "Zernike polynomials and atmospheric turbulence," *J. Opt. Soc. Am.*, **66**, No. 3, pp.207–211, 1976.
43. D.E. Novoseller, "Zernike-ordered adaptive-optics correction of thermal blooming," *J. Opt. Soc. Am. A*, **5**, pp. 1937–1942, 1988.
44. P.H. Hu, J. Stone, T. Stanley, "Application of Zernike polynomials to atmospheric propagation problems," *J. Opt. Soc. Am. A*, **6**, pp. 1595–1608, 1989.
45. N. Roddier, "Atmospheric wavefront simulation using Zernike polynomials," *Opt. Eng.*, **29**, No. 10, pp. 1174–1180, 1990.
46. M.S. Belen'ki, V.P. Lukin, V.L. Mironov, V.V. Pokasov, *Coherence of Laser Radiation in the Atmosphere*, Nauka, Novosibirsk, 1985.
47. N.N. Kalitkin, *Numerical Methods*, Nauka, Moscow, 1978.

48. V.I. Klyatskin, *Statistical Description of Dynamical Systems with Fluctuated Parameters*, Nauka, Moscow, 1975.
49. P.A. Konyaev, V.P. Lukin, V.L. Mironov, "Efficient application of adaptive optical systems in self-action conditions in the atmosphere," *Izv. Akad. Nauk SSR. Ser. Fiz.*, **49**, pp. 536–540, 1985.
50. P.A. Konyaev, "Numerical solution of stochastic problems by splitting method," in *Abstracts of Reports at the VII Conference on Laser Beam Propagation in the Atmosphere*, pp.104–106, 1983.
51. L.C. Bradley, J. Herrmann, "Numerical calculation of light propagation in a nonlinear medium (abstract)," *J. Opt. Soc. Am.*, **61**, No. 5, pp. 668, 1971.
52. J. Wallace, "Effects of nonlinear refraction at 10.6 m km on the far-field irradiance distribution," *J. Opt. Soc. Am.*, **62**, No. 3, pp. 373–378, 1972.
53. V.I. Talanov, "Focusing of light in cubic media," *Pis'ma JETF*, **11**, pp. 303–305, 1970.
54. E.V. Ustinov, "Application of Talanov's coordinates transformation for calculation of a thermal blooming optical beam in a gas jet," *Izv. V.U.Z. Radiofiz.*, **34**, pp. 40–46, 1991.
55. J.Y. Wang, L.K. Markey, "Modal compensation of atmospheric turbulence phase distortion," *J. Opt. Soc. Am.*, **68**, pp. 78–87, 1978.
56. V.S. Komarov, G. Ya. Kardasheva et al., in: *Third All-Union Conference on Atmospheric Optics and Actinometry*, Part 2, Abstracts, pp. 227–229, 1983.
57. V.S. Komarov, A.A. Mitcel', I.I. Ippolitov, T.V. Blakhovskaya, in *Seventh All-Union Symposium on Laser and Acoustic Sounding of the Atmosphere*, Part 2, Abstracts, IAO, Siberian Branch of the Russian Academy of Sciences, Tomsk, pp. 248-251, 1982.
58. V.S. Komarov, ed., *Statistical Characteristics of Temperature and Humidity Fields in the Atmosphere of the Northern Hemisphere*, Part 4, *Handbook*, Gidrometeoizdat, Leningrad, 1981.
59. Yu.S. Makushkin, A.A. Mitcel', L.I. Nesmelova, et al., *Automated System for Calculation of the Absorption Characteristics of the Atmosphere,* Part 1, *Izv. V.U.Z. Ser. Fiz.*, 1984, Manuscript on deposit with VINITI.

CHAPTER 2

Modeling an Adaptive Optics System

In Chapter 1, we discussed methods for numerically simulating the propagation of an optical wave through the atmosphere. The approaches considered allow adequate modeling of turbulent distortion and thermal blooming of coherent optical radiation. However, the development of a numerical model for atmospheric distortions of optical waves is not our objective; it is only a tool for studying the efficiency of adaptive correction of distortions introduced by a propagation medium in the operation of optical systems.

In the schemes and algorithms of adaptive correction discussed in this book, it was presumed that the optical feedback loop necessarily included three principal elements: a reference wave containing information on inhomogeneities of the refractive index of the medium, a wavefront sensor extracting this information, and a wavefront corrector. In a system operating by the phase conjugation algorithm, the corrector introduces predistortions into the emitted wave, and in the compensation system it corrects aberrations of the radiation received.

To study the efficiency of adaptive correction of atmospheric distortions, one should keep in mind the finite spatial and temporal resolution of an adaptive system, i.e., its ability to control the wavefront (or wave phase) to be corrected with a certain speed and in some finite range of spatial scales. The temporal resolution of an adaptive system is determined on the one hand by the correction algorithm and on the other hand by the operating band of the frequencies of the electronic, mechanical, and optical elements of the system. The spatial resolution, in turn, is mostly determined by the geometry of such key elements of the system as the wavefront sensor and the corrector. In this chapter we consider just this aspect of the problem, i.e., development of an efficient numerical model of the sensor and corrector from the viewpoint of making allowance for their geometry, in particular the arrangement and number of elements.

An important factor that affects the efficiency of adaptive optics systems using a reference wave is the technique by which the source of this wave is formed, as well as the principles and methodology of extracting information on atmospheric distortions from this wave. Recently, some new aspects to this problem have arisen. One of them is the description of phase distortions of the wave with wavefront dislocations that are due to points with zero intensity arising in the reference wave [1, 2]. Another no less interesting and urgent aspect is the use of an artificial reference star [3, 4] formed using the effect of laser radiation scattering in the atmosphere.

Thus, in this chapter we consider the problems of theoretical analysis and numerical simulation of the following principal elements constituting adaptive optics systems: (1) the source of the reference wave, (2) a sensor of wavefront distortions, and (3) a wavefront corrector. Here we also discuss the problems of efficient and adequate simulation of the following versions of AOS elements: an independent reference source, a natural reference star, an artificial reference star, an ideal quadrature phase sensor, an ideal phase difference sensor, a local tilt sensor (Hartmann sensor), a modal corrector, a deformable mirror with given response functions, and a segmented mirror consisting of hexagonal elements. This study is necessary for conducting the numerical experiments on adaptive correction of atmospheric distortions described in Chapters 3 and 4 of this book. Every section of this chapter considers one of the AOS elements listed here.

2.1 A Reference Wave in an Adaptive Optics System

Most adaptive systems use a reference wave to obtain information on atmospheric aberrations [5, 6]. The almost the sole exception is the system of a priori correction [7, 8]. In phase conjugation systems [9-11], a flare from a target, an additional laser, or radiation backscattered by the atmosphere serves as a source of the reference wave [3]. In image correction systems employing compensation for wavefront aberrations [12, 13], a part of the energy of the wave to be corrected is used to measure distortions.

To estimate the limiting capabilities of adaptive correction, we consider the reference wave to be coherent monochromatic radiation described by the complex amplitude $U(\vec{\rho}, z, t)$,

$$\vec{E}_{\text{ref}}(\vec{\rho}, z, t) = U(\vec{\rho}, z, t)\exp(i\omega t + ikz). \tag{2.1.1}$$

In a paraxial approximation, propagation of this wave in the negative direction of the z-axis is described by the following parabolic equation:

$$-2ik\frac{\partial U}{\partial z} = \left[\frac{\partial^2}{\partial x^2} + \frac{\partial^2}{\partial y^2} + k^2\left(n^2/n_0^2 - 1\right)\right]U. \tag{2.1.2}$$

In image correction systems, the wave to be corrected propagates in the same direction as the reference wave and is described similarly, whereas in beam-focusing systems the propagation directions of these waves are opposite; i.e., the complex amplitude of the beam to be corrected,

$$\vec{E}(\vec{\rho}, z, t) = \vec{e} \times E(\vec{\rho}, z, t) \times \exp(i\omega t - ikz), \tag{2.1.3}$$

is described in the paraxial approximation by the following equation:

$$2ik\frac{\partial E}{\partial z}=\left[\frac{\partial^2}{\partial x^2}+\frac{\partial^2}{\partial y^2}+k^2\left(n^2/n_0^2-1\right)\right]E. \qquad (2.1.4)$$

Obviously, for a closed description of an adaptive loop, Eq. (2.1.2) should be complemented by the corresponding boundary condition in the plane of the reference wave source.

2.1.1 Counterpropagating a reference beam in a phase conjugation system

Consider a system operating by the phase conjugation algorithm. To study the limiting capabilities of this algorithm, it is logical to set the reference wave so that in the absence of distortions the beam propagates in the same way it does in the absence of adaptive control. To do this, the boundary condition for the field E_0 of the corrected beam should be written through the deviation ΔS of the phase S of the reference wave from the diffraction distribution S_d:

$$E_0(\vec{\rho})=E(0,\vec{\rho})=A_0(\vec{\rho})\exp(i\Delta S),\ \ \Delta S=S(\vec{\rho})-S_d(\vec{\rho}), \qquad (2.1.5)$$

or, what is equivalent,

$$E_0(\vec{\rho})=A_0(\vec{\rho})\exp\left\{i\arg\left[U(\vec{\rho})\times U_d^*(\vec{\rho})\right]\right\}. \qquad (2.1.6)$$

Here A_0 is the complex amplitude of the initial beam in the plane of the emitting aperture. If the wavelengths of the reference beam and the beam to be corrected are different, then adaptive correction should be formulated in terms of the wavefront; i.e., through an eikonal of the reference wave θ:

$$E_0(\vec{\rho})=A_0(\vec{\rho})\times\exp\{ik[\theta(\vec{\rho})-\theta_d(\vec{\rho})]\}. \qquad (2.1.7)$$

Equations (2.1.6) and (2.1.7) impose no restrictions on the reference wave, but they assume that the diffraction-limited phase of the reference wave is known.

We can proceed from the fact that the phase conjugation correction is usually formulated as a wavefront reversal (WFR):

$$E_0(\vec{\rho})=U_0^*(\vec{\rho}) \quad \text{(for WFR)} \qquad (2.1.8)$$

in the absence of amplitude control; i.e.,

$$E_0(\vec{\rho}) = |A_0(\vec{\rho})| \exp\{-i \arg[U_0(\vec{\rho})]\} \quad \text{(for PC correction)}. \qquad (2.1.9)$$

For the result of correction by Eq. (2.1.9) at diffraction propagation to be the same that for an uncorrected beam, it is sufficient to impose the following condition on the field of the reference wave:

$$U_0^d(\vec{\rho}) = \text{const}\, A_0(\vec{\rho});\ \text{i.e.,}\ \arg\left[U_0^d(\vec{\rho})\right] = \arg[A_0(\vec{\rho})]. \qquad (2.1.10)$$

Because of reversibility of the wave equation, this condition is fulfilled if the boundary condition for the reference wave is expressed through the diffraction solution for the uncorrected beam:

$$U(\vec{\rho}, L) = \text{const}\, E^d(\vec{\rho}, L), \qquad (2.1.11)$$

where L is the length of the propagation path.

In such a formulation of the boundary condition for the reference wave, the PC moves toward the exact WFR as optical distortions on the path decrease. This guarantees that at least for small distortions, the result of correction with boundary condition (2.1.11) is closer to the diffraction solution than with any other boundary condition. Thus, the estimates of the correction efficiency can be considered to be estimates of the limiting efficiency (if the efficiency criterion is chosen so that its value reaches the maximum at diffraction-limited propagation).

As distortions increase, such a reference wave differs increasingly more from the optimal one. The problem of finding the optimal boundary conditions for the reference beam is not considered here because in practice it is usually impossible to create a reference wave with given boundary conditions.

2.1.2 Guide star in an adaptive telescope

Consider now the problem of modeling a reference source in an adaptive image correction system. In this case, the reference wave and the wave to be corrected propagate in the same direction. In a particular case in which a part of the energy of the wave to be corrected is directed onto the sensor of wavefront distortions, this wave also serves as a reference wave at the same time.

In an image correction system such as a ground-based astronomical adaptive telescope, the source of the reference wave is either a natural or an artificial star. (We will return to this problem in Chapter 5). For a natural star, the main sources of the residual error of correction are angular anisoplanatism [14-18] and quantum noise connected with low radiation intensity. For an artificial (laser) guide star, the error of correction is connected with an additional (other) type of anisoplanatism caused by the fact that the divergent reference wave and the plane wave to be corrected pass through different turbulent inhomogeneities.

Another source of error is jitter in the location of the artificial star, which makes it impossible to correct the global tilt.

It is not difficult to write the boundary conditions for a reference wave whose source is a natural star. It is a simple plane wave incident on the atmosphere at some angle to the *z*-axis:

$$U_0(\vec{\rho},\vec{\gamma}) = A\exp(ik\vec{\gamma}\vec{\rho}), \tag{2.1.12}$$

where $\vec{\gamma}$ is the vector determining the angle of inclination to the system's axis. However, if this angle is larger than 1 arcsec, we face some difficulties in numerically simulating propagation of the reference wave. Actually, the spatial spectrum of the complex amplitude of such radiation has the form

$$\tilde{U}(\vec{\kappa}) = \int_{-\infty}^{\infty}\int_{-\infty}^{\infty} A\exp(ik\vec{\gamma}\vec{\rho})\exp(i\vec{\kappa}\vec{\rho})d^2\rho = A\delta(k\vec{\gamma}) \tag{2.1.13}$$

and corresponds to the spatial frequency $\kappa = k\gamma$. The maximum spatial frequency that can be presented adequately on a grid with the step $\Delta\rho$ according to the Kotelnikov sampling theorem is equal to

$$\kappa_{\max} = \pi/\Delta\rho\,. \tag{2.1.14}$$

From the condition $k\gamma < \kappa_{\max}$ we have

$$\gamma < \pi/k\Delta\rho = \lambda/2\Delta\rho\,. \tag{2.1.15}$$

For the typical values of $\lambda = 10^{-6}$ m and $\Delta\rho = 10^{-1}$ m,

$$\gamma < 5\times10^{-6} \approx 1\,\text{arcsec}. \tag{2.1.16}$$

At the same time, to estimate the effect of angular anisoplanatism on the quality of image correction, angular distances on the order of 10 arcsec are most significant from a practical viewpoint; therefore, calculations should be performed on a grid with a step of 10^{-2} m. In this case, to make calculations for a 10-m telescope, we need a 1000×1000 grid, leading us to unjustifiably high computational expense.

This expense can be decreased in the following way: The boundary condition for the complex amplitude is formulated similarly to that of a wave propagating exactly along the axis of the optical system ($\gamma = 0$), and every phase screen $S(\vec{\rho})$ is shifted in the cross-wise direction by the distance $\vec{\Delta} = \vec{\gamma}z$ that is equal to the distance between the axis of the optical system and the "central" ray of the plane wave; i.e., the following substitution occurs:

$$U_0(\vec{\rho},\vec{\gamma}),\ S(\vec{\rho}) \Rightarrow U_0(\vec{\rho},\vec{\gamma}=0),\ S(\vec{\rho}-\vec{\gamma}z). \qquad (2.1.17)$$

For $\gamma = 10$ arcsec, the shift is roughly 1 m at the top boundary of the atmosphere z = 20 km. Note that in simulating the large-scale segment of the spectrum of atmospheric turbulence by the polynomial method, this requires the diameter of the circle of expansion to be also increased by 1 m. If the direction of the vector $\vec{\gamma}$ coincides with the direction of the vector $\vec{V}$, then the spatial shift of the phase screen can be replaced with the equivalent time interval; i.e.,

$$U_0(\vec{\rho},\vec{\gamma}),\ S(\vec{\rho},t) \Rightarrow U_0(\vec{\rho},\vec{\gamma}=0),\ S\left(\vec{\rho},\vec{\gamma}z/\vec{V}\right). \qquad (2.1.18)$$

Consider then the problem of simulation of an artificial guide star. The idea of forming an artificial reference source arose because of the need to eliminate the correction error caused by angular anisoplanatism. Therefore, the location of such a reference source almost coincides with the direction of the optical axis.

A laser guide star is formed by radiation backscattered in the atmosphere. In this case, one has to simulate both upward propagation of the laser beam and downward propagation of the backscattered radiation. Both cases are complicated by some difficulties that require modification of the numerical method.

One difficulty arises when a calculation is performed for a focused beam with a large Fresnel number. For example, for a 3-m telescope (aperture D=3 m) when laser radiation is used at a wavelength of 0.5×10^{-6} m focused at the altitude L = 10 km, the Fresnel number is

$$N_F = \frac{D}{\sqrt{\lambda L}} = \frac{3}{\sqrt{0.5\times10^{-6}\times10^4}} \approx 42, \qquad (2.1.19)$$

and the wave parameter for the Gaussian beam (for $R = D/2$) is

$$\Omega^{-1} = \frac{L}{kR^2} = \frac{\lambda L}{2\pi(D/4)^2} = \frac{0.5\times10^{-6}\times10^4}{6.28\times0.75^2} \approx 1.5\times10^{-3}. \qquad (2.1.20)$$

At such a small wave parameter, the need to apply lens conversion is obvious, but its use in such a situation is rather problematic. At the same time, the main factor giving rise to the residual error of correction is of a purely geometric character. The large Fresnel number for a relatively short vertical atmospheric path also leads to the conclusion that the approximation of geometrical optics is probably acceptable for this problem. Such an approximation is used in Fried's paper [19]. However, this paper is analytical. In our case, this geometrical optics treatment should be included in the numerical model developed for the wave optics. The following solution of this problem is proposed.

The upward propagation of the radiation is viewed as the formation of a convergent cone of rays crossing in the focal plane of a lens, and the diameter of the cone base is equal to the diameter of the emitting aperture. In simulating the upward propagation, it is important to determine the shift of the focal spot in the cross direction caused by random refraction. Diffraction and turbulent broadening can be neglected here in the first approximation because the size of the sensor subaperture d is usually selected to be roughly equal to the coherence length r_0. As a result, the angular size of the focal spot ($\sim\lambda/r_0$) roughly corresponds to the diffraction size of the spot at the focus of a subaperture of the Hartmann sensor ($\sim\lambda/d$). Therefore, the reference source is visible through the subaperture as an almost unresolved object.

The shift of the focal spot (cone vertex) can be written as

$$\Delta\vec{\rho} = \int_0^L \vec{s}(z)\,dz\,, \tag{2.1.21}$$

where L is the path length and $\vec{s}(z)$ is the vector determining the direction of the beam axis in the cross section z. In a numerical experiment, because of discrete representation of a randomly inhomogeneous medium, this integral is expressed as a sum corresponding to approximation of the integral by the method of rectangles:

$$\Delta\vec{\rho} = \sum_{j=1}^{N_Z} \vec{s}(z_j)(z_{j+1} - z_j)\,, \tag{2.1.22}$$

where z_j is the location of the jth phase screen and N_z is the number of phase screens. The tilt of the beam's axis is determined by its refraction at all previous phase screens, including the current one; i.e.,

$$\vec{s}(z_j) = \sum_{j'=1}^{j} \vec{s}_{j'}\,, \tag{2.1.23}$$

where $\vec{s}_j$ is the contribution of the jth screen. It is proposed to determine it as the average gradient of the phase S_j, with averaging performed over the cross section of the cone in the plane z_j:

$$s_j^x = \iint dx\ dy \frac{\partial S_j}{\partial x} = \int_{-r}^{r} dy\left[S_j(x_r, y) - S_j(-x_r, y)\right],$$
$$x_r = \sqrt{r^2 - y^2}, \tag{2.1.24}$$

where r is the cone radius in this cross section. The origin of the coordinate system (x, y) in the integral is brought into coincidence with the location of the cone axis in the cross section.

The next step is simulation of the downward propagation of the reference wave. Assuming the reference source to be pointlike (with location $\vec{\rho}_b$) and neglecting ray bending at propagation through the turbulent medium, we can write fluctuations of the optical path for the ray coming at the point $(\vec{\rho}, 0)$ as follows:

$$\tilde{l}(\vec{\rho}) = -\vec{\rho}\frac{\vec{\rho}_b}{L} + \int_0^L \tilde{n}[\vec{\rho} + (\vec{\rho}_b - \vec{\rho})\, z/L, z]dz. \tag{2.1.25}$$

This equation has been derived under the condition

$$(\vec{\rho} - \vec{\rho}_b)^2 / L^2 \ll 1; \tag{2.1.26}$$

i.e., in the paraxial approximation, where $\vec{\rho}_b$ is the coordinate of the reference source (beacon).

The well-known conclusion that an adaptive system operating against an artificial guide star cannot compensate for image jitter follows directly from the above equation. In reality, an inhomogeneity in the form of a thin prism is situated immediately in front of the telescope aperture; i.e.,

$$\tilde{n}(\vec{\rho}, z) = \vec{\alpha}\vec{\rho}\,\delta(z), \tag{2.1.27}$$

where $\vec{\alpha}$ is the prism angle, and L is the height (distance) at which the reference source is located. From purely geometric reasoning, it is clear that the reference source formed through this lens is situated at the point

$$\vec{\rho}_b = \vec{\alpha}L. \tag{2.1.28}$$

Then, as expected,

$$\tilde{l}(\vec{\rho}) = -\vec{\rho}\frac{\vec{\alpha}L}{L} + \int_0^L \vec{\alpha}\vec{\rho}\,\delta(z)dz = -\vec{\alpha}\vec{\rho} + \vec{\alpha}\vec{\rho} = 0. \tag{2.1.29}$$

Equation (2.1.25) is basic for numerical simulation. The randomly inhomogeneous field of the refractive index is represented, according to the splitting technique, as a series of thin phase screens; this is equivalent to the following representation of $\tilde{n}(\vec{\rho}, z)$:

$$\tilde{n}(\vec{\rho}, z) = \sum_{j=1}^{N_Z} S_j(\vec{\rho})\delta(z_j). \tag{2.1.30}$$

Substituting Eq. (2.1.30) into Eq. (2.1.25), we obtain

$$\tilde{l}(\vec{\rho}) = -\vec{\rho}\frac{\vec{\rho}_b}{L} + \sum_{j=1}^{N_Z} S_j\left[\vec{\rho} + (\vec{\rho}_b - \vec{\rho})z_j/L\right]. \tag{2.1.31}$$

In numerical simulation, the cross coordinates are also discrete; that is, the values of phase distortions are known only at the nodes of the grid:

$$S_{l,m,j} = S(x_l, y_m, z_j). \tag{2.1.32}$$

In the case being considered, these values are to be interpolated to an arbitrary point. Simple two-dimensional linear interpolation seems to be sufficient for this problem. The interpolation is performed according to the following equations:

$$\begin{aligned} S(x,y) &= S(x,y_m) + \frac{S(x,y_{m+1}) - S(x,y_m)}{\Delta y}(y - y_m); \\ S(x,y_m) &= S(x_l,y_m) + \frac{S(x_{l+1},y_m) - S(x_l,y_m)}{\Delta x}(x - x_l). \end{aligned} \tag{2.1.33}$$

Here the subscripts (l, m) correspond to the left bottom node of the grid cell containing the point with the coordinates $\vec{\rho} = (x, y)$. To calculate the optical propagation difference in the plane of the emitting aperture at the nodes of the computational grid, we use Eq. (2.1.31) in the form

$$\tilde{l}_{l,m} = \tilde{l}(\vec{\rho}_{l,m}) = -\vec{\rho}_{l,m}\frac{\vec{\rho}_b}{L} + \sum_{j=1}^{N_z} S_j\left[\vec{\rho}_{l,m} + (\vec{\rho}_b - \vec{\rho}_{l,m})z_j/L\right], \tag{2.1.34}$$

with interpolation according to Eq. (2.1.33).

Thus, in this section we have considered three possible models of the reference wave. The model of a counterpropagating beam has been formulated so that the effects of anisoplanatism are minimal; i.e., the trajectories of the reference beam and the beam to be corrected coincide exactly in the absence of distortions. Therefore, when using this model, we are studying the limiting efficiency of phase correction in the pure form. The effects of anisoplanatism for such problems were considered in Ref. [20], which studied correction of turbulent distortions in a wide beam using a narrow reference beam and vice versa.

Since for vertical paths the effects of diffraction at turbulent inhomogeneities can usually be neglected, the efficiencies of the phase and amplitude-phase correction in telescopes differ insignificantly. Therefore, anisoplanatism is the main factor in the problems of adaptive formation of images of astronomical objects. Anisoplanatism is a purely geometric factor, and in this respect the models of a guide star discussed here are different than the model of a counterpropagating beam.

2.2 Wavefront Sensors

In Sec. 2.1, we considered the problem of numerical simulation of a reference source that emits a wave bearing information on inhomogeneities in the refractive index field of the propagation medium. To extract this information for further use in control problems, some electro-optical device should be used. Such devices in general are called "wavefront sensors." This section is devoted to their numerical simulation.

In a general form, the problem of simulating a wavefront sensor can be formulated in the following way: there is a grid representation of the complex amplitude of an optical wave

$$U_{I,J} = U(\vec{\rho}_{I,J}),\ I,J = 0,\ 1,\ldots,N-1 \tag{2.2.1}$$

resulting from numerical solution of the propagation problem. To be determined is a two-dimensional (grid or continuous) function, which is an estimate of the phase φ (or eikonal θ) of the wave, and the accuracy of this estimate characterizes the spatial and temporal resolution of the wavefront sensor to be simulated. For an ideal sensor, this estimate should differ from the true value by no more than a constant. It is also desirable for the estimate to be a continuous function, since the main types of wavefront correctors are incapable of reproducing a discontinuous surface.

2.2.1 Ideal quadrature sensors

Some problems of adaptive optics involve determining the limiting efficiency of phase correction algorithms as compared with algorithms of amplitude-phase correction. Toward this end, it is sufficient in principle to have a model of some ideal sensor. Because in the numerical experiment we deal with both the real part of the complex amplitude Re$E = A\cos\varphi$ and the imaginary part of the complex amplitude Im$E = A\sin\varphi$, such an ideal sensor can be called the "quadrature sensor" [21]. The phase at every point (node of the computational grid) is calculated as the principal argument of the complex amplitude arg(E). Since the area of arg(E) is limited by the range $[0, 2\pi]$, lines of phase discontinuity inevitably arise for aberrations larger than the wavelength λ. If these lines are closed, they can be removed by adding or subtracting $2\pi n$ (n is an integer number) in the area enclosed by the discontinuity.

Direct measurement of the quadrature components of electromagnetic oscillations in the optical region is practically impossible. The operating principle of any wavefront sensor is based on intensity measurements. Certainly, there are a number of wavefront sensor alternatives to the Hartmann sensor [22]. Among them are the wavefront curvature sensor and phase contrast sensors [23-25]. A wave is first subjected to some (diffraction or interference) transformation, and then the intensity distribution pattern is analyzed mathematically [26]. The result of this analysis is the estimate of the 2D phase distribution that is sought.

2.2.2 Ideal phase difference sensor

There are two main approaches to the problem of wavefront measurement: one is based on measurement of the phase difference, and the other on measurement of the local tilt. Measurement of phase difference is based on the interference transformation, and the optical part of the sensor is an interferometer. Adaptive systems mostly employ lateral shear interferometers. The result of measurement in a shear interferometer is an estimate of the phase difference between two small areas, and in the ideal case it is the phase difference between two spatially separated points.

Using numerical simulation of optical radiation propagation, it is possible to form, in a natural way, a two-dimensional array of phase differences between neighboring nodes of the computational grid in the directions of the 0x- and 0y-axes. The elements of this array can be written as

$$\begin{aligned} \Delta^x_{I,J} &= \arg\left(U_{I+1,J}U^*_{I,J}\right), \\ \Delta^y_{I,J} &= \arg\left(U_{I,J+1}U^*_{I,J}\right), \end{aligned} \tag{2.2.2}$$

respectively. These differences can be distorted by imposing, for example, additive noise that imitates errors of the interferometric sensor:

$$\begin{aligned} \Delta^x_{I,J} &= \arg\left(U_{I+1,J}U^*_{I,J}\right) + \mu^x_{I,J}; \\ \Delta^y_{I,J} &= \arg\left(U_{I,J+1}U^*_{I,J}\right) + \mu^y_{I,J}. \end{aligned} \tag{2.2.3}$$

The reconstruction of the phase as applied to problems of adaptive optics was considered in sufficient detail in Refs. [27]–[31]. In numerical simulation, the direct method based on the use of a two-dimensional discrete Fourier transform is most convenient [32-34] because the complex amplitude of the field is set on a uniform grid whose dimensions allows application of a fast Fourier transform (FFT). The number of values of the phase difference is twice as large as the number of points at which the phase is sought; therefore, the problem is too large to solve. Thus, additional conditions are usually imposed on the problem, namely, minimization of the square discrepancy [35]:

$$\sum_{I,J}\left[\left(\varphi_{I+1,J}-\varphi_{I,J}\right)-\Delta^{x}_{I,J}\right]^2+\left[\left(\varphi_{I,J+1}-\varphi_{I,J}\right)-\Delta^{y}_{I,J}\right]^2 \rightarrow \min, \quad (2.2.4)$$

or minimization of the integral variance of the estimation error [28]:

$$\left\langle \sum_{I,J}\left(\varphi_{I,J}-\hat{\varphi}_{I,J}\right)^2 \right\rangle \rightarrow \min. \quad (2.2.5)$$

Here the angle brackets indicate statistical averaging over the noise ensemble, $\varphi_{I,J}$ is the sought estimate of the phase, and $\hat{\varphi}_{I,J}$ is the exact value of the phase. In both cases, the problem is reduced to the solution of a set of linear equations in the following form:

$$\begin{aligned}&\varphi_{I+1,J}+\varphi_{I-1,J}+\varphi_{J+1}+\varphi_{J-1}-4\varphi_{I,J}\\&=\Delta^{x}_{I,J}+\Delta^{y}_{I,J}-\Delta^{y}_{I,J-1}-\Delta^{x}_{I-1,J}=f_{I,J}.\end{aligned} \quad (2.2.6)$$

The solution of this set of equations is based on the representation of the sought and known functions as a Fourier series and the solution has the following form:

$$\Phi_{L,M}=\frac{F_{L,M}}{2\left[\cos\left(2\pi L/N\right)+\cos\left(2\pi M/N\right)-2\right]}, \quad (2.2.7)$$

where Φ is a two-dimensional Fourier transform of the phase φ array, and F is two-dimensional Fourier transform of the right-hand side of the initial equation. If the denominator vanishes at $L = M = 0$, this indicates that the problem can be solved accurately to a constant.

Since the values of the sought function (phase) at the boundaries are not known, the boundary conditions should be given for the phase difference. These boundary conditions are included in the right-hand side of the set of equations in (2.2.6), but they are connected to the values of the phase at the points lying beyond the grid area:

$$\Delta^x_{N-1,J} = \varphi_{N,J} - \varphi_{N-1,J};\ \Delta^x_{-1,J} = \varphi_{0,J} - \varphi_{-1,J},\quad J = 0,\dots,N-1;$$
$$\Delta^y_{I,N-1} = \varphi_{I,N} - \varphi_{I,N-1};\ \Delta^y_{I,-1} = \varphi_{I,0} - \varphi_{I,-1},\quad I = 0,\dots N-1. \tag{2.2.8}$$

Thus, assignment of the boundary conditions depends on the method of extrapolating the sought function beyond the domain of definition. Any grid function given at a finite $N \times N$ grid can be complemented with the periodic one by at least one of the two following methods. In the first case, the grid function is directly extrapolated periodically without the application of any transformation:

$$\varphi_{I-N,J} = \varphi_{I,J} = \varphi_{I+N,J};\quad \varphi_{I,J-N} = \varphi_{I,J} = \varphi_{I,J+N}, \tag{2.2.9}$$

and a discontinuity (jump) is possible at the boundaries of the periods. The differences at the boundaries can be calculated through the known values in the following way:

$$\Delta^x_{-1,J} = \Delta^x_{N-1,J} = \sum_{I=0}^{N-2} \Delta^x_{I,J};\ \Delta^y_{I,-1} = \Delta^y_{I,N-1} = \sum_{J=0}^{N-2} \Delta^y_{I,J}. \tag{2.2.10}$$

The other method is to construct a new grid function on the double-sized grid by a mirror reflection of the initial grid function [33, 36]. The new grid function obtained in such a way continues periodically in both directions with a period of $2N$. The differences at the boundaries of the period for such a function are equal to zero, and the problem of reconstruction is solved by the same equation (2.2.6), the only difference being that the grid size N is replaced by the double value $2N$. In the solution obtained, only a quarter of the values corresponding to the initial N-dimensional grid are used.

Since the same set of equations is solved in both cases, the solution apparently is independent of the method of periodic extrapolation, at least in the absence of noise. Obviously, the first method is more economical in the number of mathematical operations because the discrete Fourier transform is performed on the $N \times N$ matrix, whereas in the second case the matrix is $2N \times 2N$.

Thus, solution of the set of equations in (2.2.6) with the right-hand side given by Eq. (2.2.2) allows the phase (argument) of the complex amplitude to be reconstructed accurately to a constant. The addition of noise by Eq. (2.2.3) imitates measurement errors. As a result, this mathematical model corresponds to an interferometric sensor with a spatial resolution equal to the distance between the grid nodes.

This algorithm also provides a solution to the phase joining problem that arises in the model of an ideal quadrature sensor, but only if all discontinuity lines of the phase function are closed. This algorithm can be applied instead of constructing an algorithm that seeks discontinuity lines; in addition, it automatically performs joining. It should be kept in mind that all discontinuities, including unclosed ones that join phase dislocations, are eliminated in this case. Thus, the algorithm for solving the problem of reconstructing the grid function from the known values of its first differences can be used both for simulation of an ideal sensor and for approximate simulation of an interferometric sensor.

2.2.3 Hartmann sensor

Consider the problem of simulation of the second type of sensor: the Shack–Hartmann sensor. The idea that forms the basis for operation of these sensors was formulated by Hartmann [37]. It consists of measurement of local tilts of the wavefront at the sensor's subapertures, followed by calculation of the tilt map for the entire aperture. Initially this method was used for testing large-sized (mostly astronomical) optics, i.e., as a method of optical testing [38].

In the early 1970s, the technology to produce the special Hartmann-type diaphragm was developed [39]. This has made it possible to use the Hartmann sensor in adaptive optics systems for measuring atmospheric aberrations of the wavefront in real time. This modified sensor is also called the Shack–Hartmann sensor.

The Shack–Hartmann sensor is now used most widely in adaptive optics systems. The basic layout of this sensor is shown in Fig. 2.2.1. The sensor consists of an array of collecting lenslets in whose focal plane the photodetector array is situated. Usually the lenslets (subapertures) are arranged in rows, but sometimes other configurations (e.g., hexagonal packing) are used [40]. To amplify the light flux incident on the photodetectors, photomultiplier tubes or other amplifiers are applied.

The operating principle of the Shack–Hartmann sensor consists of is the following: Every subaperture focuses the corresponding partial beam onto the plane of the photodetectors. The signals coming from the photodetectors are used to estimate the shift of every beam from the optical axis of the corresponding lenslet. This shift is assumed to be proportional to the mean local tilt of the wavefront within the corresponding subaperture. If an array with a sufficiently large number of photosensitive elements [e.g., a charge-coupled device (CCD) array] is used, the shift of the focal spot is determined as a shift of its centroid. Otherwise, redistribution of energy between quadrants of a four-quadrant photodetector is recorded, and the shift is estimated from the predetermined direction-finding characteristic. In some systems, the four-quadrant detector operates in the zero stabilization mode [40]; i.e., the orientation of the detector is adjusted until the image of the light spot creates an equal light level on all four quadrants. For numerical simulation of the Hartmann sensor, we have implemented two approaches [41].

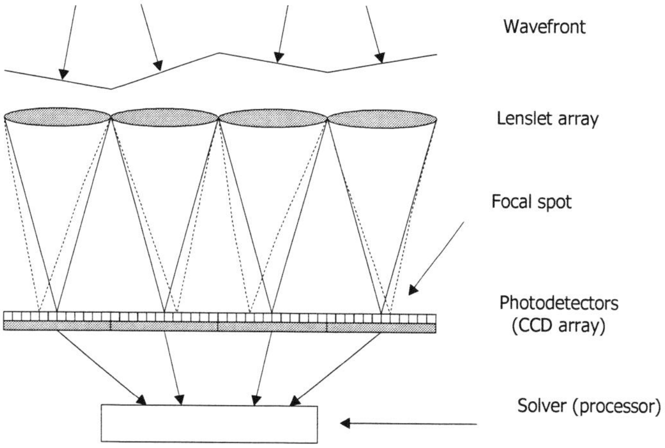

Figure 2.2.1. Basic layout of the Shack–Hartmann sensor.

2.2.3.1 First method for modeling the Hartmann sensor

This method is based on direct calculation of the intensity distribution in the focal spot for every subaperture following estimation of the shift of its centroid. Toward this end, fragments corresponding to subapertures are separated from the initial array of the complex amplitude. The two-dimensional arrays obtained usually have a small size; therefore it makes sense to complement them with zeros to obtain the intensity distribution at the focus on a denser grid. The input pupil of the sensor has the shape of a circle, and therefore the single-amplitude mask should be placed on the distribution of the complex amplitude. The amplitude mask is also placed on every subaperture. The hyper-Gaussian mask smoothes the sharp edges of a subaperture and improves the accuracy of solution of the diffraction problem by preventing overlapping in the region of spatial frequencies.

The next step of the algorithm is calculation of the field intensity distribution at a subaperture focus. This step can be carried out in either of two ways: by solving the problem of diffraction of a focused beam or by calculating the angular distribution of the field in the far zone. In both cases, the result is practically the same, since in the paraxial approximation the intensity in the focal plane coincides with the angular spectrum of the field. The difference is only in the grid scale factor.

An important factor affecting the measurement error is shot noise caused by the quantum nature of light and the photodetector dark current. In numerical simulation of the sensor, it is convenient to set the mean number of photons $\langle N_p \rangle$ falling on the sensor aperture during the exposure with allowance for the quantum efficiency of photoelements as an input parameter. Every node of the computational grid in the plane of the focal spot corresponds to a small area of the photodetector. For every such area, the mathematical expectation of the number of photoelectrons $\langle n_p \rangle$ is determined as a product of $\langle N_p \rangle$ and the fraction of light energy incident on this area. From the calculated mean $\langle n_p \rangle$, the corresponding random value n_p is generated, which obeys Poisson statistics. The shift of the focal spot is estimated as deviation $\Delta\vec{\rho}$ of the centroid of the distribution of grid realization $n_p(x_I, y_J)$. Then the local tilt is estimated as a value proportional to the centroid shift and inversely proportional to the focal length f:

$$\vec{s} = \Delta\vec{\rho}_c / f \,. \tag{2.2.11}$$

2.2.3.2 Second method for modeling the Hartmann sensor

This modeling method allows the calculation of the intensity distribution at the focus of every subaperture to be omitted. It is based on direct calculation of the centroid shift as a gradient of the phase φ weighted to the intensity I and averaged over the subaperture A. Starting from the parabolic equation, one can readily derive the well-known equation [42] for the vector $\vec{\rho}_c$ of the focal spot centroid within the subaperture A:

$$\vec{\rho}_c = \frac{kf}{P} \iint_A I(\vec{\rho})\ \vec{\nabla}\varphi\ d^2\rho \,. \tag{2.2.12}$$

Here P is the total power of radiation incident on a subaperture. However, the use of this equation is complicated by a difficulty connected with differentiation of the phase. If after solution of the propagation problem we know the complex amplitude, then there is no guarantee that we will succeed in obtaining the continuous phase. To avoid this, the weighted phase gradient should be expressed through the gradients of the real and imaginary parts of the complex amplitude E:

$$\begin{aligned} I(\vec{\rho})\nabla\varphi &= |E|^2\,\vec{\nabla}\mathrm{arctg}(\mathrm{Im}\,E/\mathrm{Re}\,E) \\ &= (\mathrm{Re}\,E)\vec{\nabla}(\mathrm{Im}\,E) - (\mathrm{Im}\,E)\vec{\nabla}(\mathrm{Re}\,E). \end{aligned} \tag{2.2.13}$$

Then, we have the following equation for estimating the local tilt:

$$\vec{s} = \frac{1}{P}\iint_A \left[(\mathrm{Re}\,E)\vec{\nabla}(\mathrm{Im}\,E) - (\mathrm{Im}\,E)\vec{\nabla}(\mathrm{Re}\,E)\right] d^2\rho \,. \tag{2.2.14}$$

This equation can also be derived more rigorously by starting from the parabolic equation for the complex amplitude without invoking the concept of phase. In the computer model, the partial derivatives of the real and imaginary parts of the field are calculated with the use of the discrete Fourier transform, which is performed at once on the entire grid, and then integrals for separate subapertures are estimated. This significantly reduces the number of mathematical operations needed to simulate a sensor with a large number of subapertures.

Reconstruction of the wavefront can also be carried out by different methods. There are two main approaches. In one case, local tilts are used to calculate phase differences and then algorithms similar to those in interferometric sensors are applied [28-30]. In another approach, the wavefront is represented as a sum of some functions, and then the local tilts of the wavefront are fitted to the measured values.

For modal reconstruction of the wavefront on the sensor aperture from estimates of local tilts, we have implemented a method [43, 44] in which the sought distribution is presented as an expansion in some Zernike polynomials Z_l:

$$\varphi(\vec{\rho}) = \sum_{l=1}^{N_Z} a_l Z_l(\vec{\rho}) \tag{2.2.15}$$

(a_l are the expansion coefficients) and local tilts in this expansion are fitted to the estimates obtained in the wavefront sensor:

$$\sum_{m=1}^{M} \left[\sum_{l=1}^{N_Z} a_l \vec{Z}_{lm} - \vec{s}_m \right]^2 \to \min . \tag{2.2.16}$$

Here, $\vec{Z}_{lm}$ is the local tilt of the lth basis function Z_l on the mth subaperture, and $\vec{s}_m$ is the tilt measured on the mth subaperture. The local tilt Z_{lm} can be determined by different methods; for example, as a gradient at the center of the subaperture [43] or a gradient averaged over the subaperture area [44]. It is easiest to determine the components of the local tilt as coefficients of approximation by a linear function; i.e., as a result of fitting to the closest plane. The minimization problem [Eq. (2.2.16)] is solved by the ordinary variational method, i.e., the partial derivatives with respect to the expansion coefficient are set at zero:

$$\frac{\partial}{\partial a_k} \sum_{m=1}^{M} \left[\sum_{l=1}^{L} a_l \vec{Z}_{lm} - \vec{s}_m \right]^2 = 0 , \tag{2.2.17}$$

and the resulting set of linear equations can be written in a matrix form as

$$\left\| A_{kl} \right\| \left\| a_l \right\| = \left\| B_k \right\| , \tag{2.2.18}$$

where B_k is the right-hand side of the kth equation, A_{kl} is the matrix of coefficients, and l is the number of the unknown a_l,

$$A_{kl} = \left(\sum_{m=1}^{M} \vec{Z}_{lm} \vec{Z}_{km} \right); \; B_k = \sum_{m=1}^{M} \vec{s}_m \, \vec{Z}_{km} \, . \tag{2.2.19}$$

The solution of this set can be represented by the inverse matrix:

$$\|a_l\| = \|A_{kl}\|^{-1} \|B_k\| . \tag{2.2.20}$$

The matrix is inverted numerically at initialization of the sensor data. Once calculated, the inverse matrix is then used at every exposure. The same occurs in the actual adaptive system; the reconstruction matrix is saved in the memory of a specialized processor whose input data are the vector of measurements.

Different sets of functions are used for the expansion in Eq. (2.2.15). Most often they are Zernike polynomials, as well as the Karhunen–Loeve expansion; however, it is worth using the latter if the statistics of aberrations are known [45]. In the case of nonlinear distortions, the Karhunen–Loeve expansion loses the sense. The software used in the model described here employs Zernike polynomials as a basis.

Another method of phase reconstruction implemented with the same software is based on a zonal principle. The first step is estimation of the phase difference Δ between the centers of neighboring subapertures:

$$\Delta_{km} = \tfrac{1}{2}(\vec{g}_k + \vec{g}_m)(\vec{\rho}_k - \vec{\rho}_m) \, ,$$

where $\vec{\rho}_k$ and $\vec{\rho}_m$ are the coordinates of the centers of the mth and kth subapertures and $\vec{g}_k$ and $\vec{g}_m$ are the estimates of the phase front tilt on these subapertures. The vector character of this record allows the algorithm to be generalized for most geometries of subapertures. In particular, we have implemented a model of a cellular configuration. For this model, the problem of reconstructing the phase corresponding to the minimal discrepancy is written as

$$\sum_k \sum_1^{\leq 6} (\varphi_k - \varphi_m - \Delta_{km})^2 = \min(\varphi_1, \ldots, \varphi_N) \, .$$

The inner sum on the left-hand side of this equation consists of six subapertures next to the kth subaperture. The minimization problem is solved by the variational method, leading to a set of linear equations whose number is equal to the number of subapertures N. At $N < 100$, inversion of the corresponding matrix creates no difficulties. Because the phase is determined accurately to some constant, an additional condition, e.g., $\varphi_1 = 0$, should be imposed on the matrix for conditionality.

The numerical model of the sensor was tested in the following way: The reference phase was set as a sum of the first ten Zernike polynomials and then the Shack–Hartmann sensor was modeled by the technique described here. The reconstruction error for the expansion coefficients in the parameters of the numerical experiment did not exceed 1% at wavefront aberrations reaching 1–2λ in an area equal to the size of the sensor subaperture.

2.3 Wavefront Correctors

In a certain sense, the wavefront corrector is a final element in an adaptive system. It realizes the possibilities embedded in the correction algorithm and provided by other elements. The efficiency of an adaptive optics system as a whole depends ultimately on the spatial and temporal resolution of the corrector.

In spite of significant technological progress, it is still a complicated and expensive task to produce a high-quality corrector. Therefore, it is an important preliminary to determine the requirements for spatial and temporal resolution that will provide the desired quality of correction. This determines the parameters in the development of numerical models of a wavefront corrector.

Controllable mirrors are now the main type of correctors applied in adaptive optics systems. Two types of adaptive mirrors are usually distinguished: deformable and segmented. Specific designs of these mirrors differ in the number of degrees of freedom, geometry of actuators, material, shape of segments, etc. Let us consider approaches to modeling existing correctors, starting from the most widely used theoretical model—a modal corrector.

2.3.1 Modal correctors

We have implemented numerical models of correctors of three types; two of them are mentioned above—deformable and segmented correctors. The third type—a modal corrector—is a hypothetical device whose response functions are components of some expansion basis; usually they are Zernike polynomials. Although the corresponding physical device does not actually exist, estimation of the efficiency of using this corrector has both theoretical and practical significance for several reasons. First, some deformable adaptive mirrors are developed and equipped with an interface, which makes it possible to reconstruct the set of aberrations corresponding to several first Zernike polynomials [46]. Second, the modal algorithm of wavefront reconstruction described in Sec. 2.2 gives the aberration coefficients from whose number (which is needed to achieve the required quality of correction) we can determine the requirements for the wavefront sensor. And third, the statistics of the coefficients of the Zernike expansion have been well studied for turbulent distortions of the wavefront [47-51], and in some cases they give rather simple equations for estimating the efficiency of an adaptive correction [52-56]. However, this is not true for nonlinear distortions of high-power beams.

When using a modal corrector, the corrected phase or wavefront is calculated as a difference between the initial wavefront and the truncated series of Zernike polynomials:

$$\Delta\varphi(\vec{\rho}) = \varphi(\vec{\rho}) - \sum_{l=1}^{L} a_l Z_l(\vec{\rho}). \tag{2.3.1}$$

The coefficients a_l of the series are usually sought from the minimization condition of the integral square error of approximation:

$$\int W(\vec{\rho}) \left[\phi(\vec{\rho}) - \sum_{l=1}^{L} a_l Z_l(\vec{\rho}/R) \right]^2 d^2\rho \rightarrow \min. \tag{2.3.2}$$

If the aperture function corresponds to a circle of the radius R:

$$W(\vec{\rho}) = \begin{cases} 1, \rho \le R, \\ 0, \rho > R, \end{cases} \tag{2.3.3}$$

then, because of orthogonality of the Zernike polynomials on a circle, the coefficients are determined as the expansion coefficients:

$$a_l = \left[\int_{\rho<R} \varphi(\vec{\rho}) Z_l(\vec{\rho}) d^2\rho \right] \Bigg/ \left[\int_{\rho<R} Z_l^2(\vec{\rho}) d^2\rho \right]. \tag{2.3.4}$$

Otherwise, the problem is solved by the variational method, giving a set of linear equations that can be written in a matrix as

$$\|A_{kl}\| \|a_l\| = \|B_k\|, \tag{2.3.5}$$

where

$$A_{kl} = \int W(\vec{\rho}) Z_k(\vec{\rho}/R) Z_l(\vec{\rho}/R) d^2\rho,$$

$$B_k = \int W(\vec{\rho}) Z_k(\vec{\rho}/R) \varphi(\vec{\rho}) d^2\rho. \tag{2.3.6}$$

In a numerical model, all functions are as a rule given on the grid and the integrals are replaced by the corresponding sums:

$$A_{kl} = \sum_{I,J} W(\vec{\rho}_{I,J}) Z_k(\vec{\rho}_{I,J}/R) Z_l(\vec{\rho}_{I,J}/R), \tag{2.3.7}$$

$$B_k = \sum_{I,J} W(\vec{\rho}_{I,J}) Z_k(\vec{\rho}_{I,J}/R)\varphi(\vec{\rho}_{I,J}).$$

In this case, solving Eq. (2.3.5), we minimize the parameter

$$\sigma^2 = \sum W_{I,J}\left[\varphi_{I,J} - \sum_{l=0}^{L} a_l Z_l(\vec{\rho}_{I,J}/R)\right]^2. \tag{2.3.8}$$

2.3.2 Deformable mirrors

There are various deformable mirrors that are different in both the design of their actuators and how they are fastened to the back surface of a mirror plate, adaptive mirrors with a bimorph basis [57-60] in which the distributed load is applied to the plane surface, etc. Mirrors with discretely arranged points of force (in some case, force moment) application have gained the widest acceptance [61]. Here we consider the simplest and therefore most widely used technique of mirror surface control involving the use of actuators. In theoretical papers, it is usually acceptable to describe such a deformable mirror through the given response functions. The mirror surface is described as a weighted sum:

$$S(\vec{\rho}) = \sum_{l=1}^{L} a_l f_l(\vec{\rho} - \vec{\rho}_l), \tag{2.3.9}$$

where a_l is the deflection of the mirror surface at the point of fastening of the lth actuator $\vec{\rho}_l$. The response functions f_l can be measured experimentally, calculated as a solution of the set of mechanical equations, or given, based on some reasons. In a number of papers [62, 63] it was shown that the Gaussian response function

$$f(\vec{\rho}) = \exp(-\rho^2/w^2) \tag{2.3.10}$$

agrees well with experimental response functions.The halfwidth w can be found from comparison with the experimental data. It typically ranges from 0.7 to 0.8 d, where d is the spacing between actuators.

To approximate the surface needed by a deformable mirror with the given response functions, the minimization problem [Eq. (2.3.8)] written in the following form:

$$\sigma^2 = \int W(\vec{\rho})\left[\varphi(\vec{\rho}) - \sum_{l=1}^{L} a_l f(\vec{\rho} - \vec{\rho}_l)\right]^2 d^2\rho \to \min \tag{2.3.11}$$

was solved.

To reduce computational expense, the response functions were given as the truncated Gaussian function:

$$f(\vec{\rho})=\begin{cases}\exp\left(-\rho^2/w^2\right), & \rho\le 2w\\ 0, & \rho>2w\end{cases}. \qquad (2.3.12)$$

The elements of the matrices [Eq. (2.3.5)] were calculated as follows:

$$A_{kl}=\sum_{|\vec{\rho}_{I,J}-\vec{\rho}_k|<2w} W(\vec{\rho}_{I,J})f_k(\vec{\rho}_{I,J}-\vec{\rho}_k)f_l(\vec{\rho}_{I,J}-\vec{\rho}_k);$$

$$B_k=\sum_{|\vec{\rho}_{I,J}-\vec{\rho}_k|<2w} W(\vec{\rho}_{I,J})f_k(\vec{\rho}_{I,J}-\vec{\rho}_k)\varphi(\vec{\rho}_{I,J}). \qquad (2.3.13)$$

The actuator fastening points were set as a equidistant grid whose node coordinates (x_l, y_l), $l = 1,\ldots, L$, were calculated by the following equations:

$$x_l=-R_c+[l/\sqrt{L}]\ d;$$

$$y_l=-R_c+\{l/\sqrt{L}\}\ \sqrt{L}\ d. \qquad (2.3.14)$$

The square brackets in Eq. (2.3.14) denote the integer part of a value and the braces are for the fractional part of a value; R_c is the radius of the corrector aperture.

2.3.3 Segmented correctors

The currently available designs of segmented correctors [64, 65] differ in two principal parameters: segment (element) shape and the number of their degrees of freedom. The shape of an element is usually either square or hexagonal. The number of degrees of freedom of every element varies from one to three and determines, respectively, position control (one degree of freedom), slope control (two degrees of freedom), and position and slope control (three degrees of freedom).

The corrector surface S can be described mathematically through the response functions:

$$S(\vec{\rho})=\sum_{l=1}^{L}a_l\, w(\vec{\rho}-\vec{\rho}_l)f_l(\vec{\rho}-\vec{\rho}_l), \qquad (2.3.15)$$

where the response function f_l is given in the form

$$f_l(\vec{\rho})=C_l;\ \ f_l(\vec{\rho})=A_l x+B_l y;\ \ f_l(\vec{\rho})=A_l x+B_l y+C_l, \qquad (2.3.16)$$

for one, two, and three degrees of freedom, respectively. The aperture function of a corrector element w is taken to be equal to unity inside the corrector plate and zero outside of it:

$$w(\vec{\rho}) = \begin{cases} 1, & \vec{\rho} \in A, \\ 0, & \vec{\rho} \notin A. \end{cases} \tag{2.3.17}$$

For a corrector with the square shape of elements, the aperture function can be easily set as

$$w(\vec{\rho}) = \begin{cases} 1, & |x| \le d/2 \wedge |y| \le d/2, \\ 0, & |x| > d/2 \vee |y| > d/2, \end{cases} \tag{2.3.18}$$

where d is the cross dimension of the element. It is a more complicated problem to set the aperture function and the central positions of elements for a corrector with hexagonal packing of segments. Some difficulties arise in formalization of the algorithm and its numerical implementation, especially when constructing a numerical model of a corrector with an arbitrary number of elements inscribed in an aperture of a given size.

In our software, this problem is solved in the following way: The input parameters of the procedure are the corrector aperture radius R and the number of "rings" of elements N. One ring corresponds to a 7-element corrector, two rings to a 19-element corrector, three rings to a 37-element corrector, and so on. The first step is determination of the radius r of a circle circumscribed around the element. It is calculated so that corrector segments completely fill an aperture of the radius R. Omitting the corresponding purely geometric consideration, we present here only the resulting equation [66]:

$$\begin{aligned} r &= R/3\ [(N+1)/2 - 1] \quad \text{for odd } N, \\ r &= R/\sqrt{3/4 + (3/2\,N + 1/2)^2} \quad \text{for even } N. \end{aligned} \tag{2.3.19}$$

Then the circle of rings of segments begins. For every ring, the circle begins from the segment whose center lies on the abscissa to the right of the corrector center. The coordinates of the segment centers are calculated by the following iterative formula:

$$\vec{\rho}_{m,k} = \begin{cases} m\,\vec{\Delta}_{m,k}, & \{k/m\}=0 \\ \vec{\rho}_{m,k-1}+\vec{\Delta}_{m,k}, & \{k/m\}\neq 0 \end{cases}, \quad m=1,\ldots,N, \;\; k=0,\ldots, \quad 6m-1;$$

$$\vec{\Delta}_{m,k} = \sqrt{3}\,r\left(\vec{e}_x \cos\xi_{m,k} + \vec{e}_y \sin\xi_{m,k}\right); \;\; \xi_{m,k} = \begin{cases} \dfrac{\pi k}{3m} & \{k/m\}=0, \\ \dfrac{\pi k}{3m}+\dfrac{2\pi}{3} & \{k/m\}\neq 0, \end{cases}$$

(2.3.20)

where m is the ring number, k is the number of a segment in the ring, and braces are for the fractional part of a value. Then the circle of grid points inside the segment begins. To determine whether a current point lies within the segment, the length of a line connecting the segment center and the boundary and passing through the current point (x, y) is calculated as

$$\rho_{\max}(x,y) = \frac{\sqrt{3}\,r}{2\cos\left(\xi_1 - \frac{\pi}{3}\left[\xi_1 / \frac{\pi}{3}\right] - \frac{\pi}{6}\right)}; \;\; \xi_1 = \arctan_2\left(|x|, y\right), \quad (2.3.21)$$

where $\arctan_2$ is the Fortran function. The aperture function of the segment is determined as follows:

$$w(\vec{\rho}) = \begin{cases} 1, & \rho \le \rho_{\max}(\vec{\rho}) \\ 0, & \rho > \rho_{\max}(\vec{\rho}) \end{cases}; \;\; \vec{\rho} = (x,y). \quad (2.3.22)$$

For every segment, the control vector $\{A_l, B_l, C_l\}$ is determined by solving the problem of minimization of the square error of the approximation of phase φ:

$$\sigma_l^2 = \int W(\vec{\rho})\,w(\vec{\rho}-\vec{\rho}_l)\left[f_l(\vec{\rho}-\vec{\rho}_l) - \phi(\vec{\rho})\right]^2 d^2\rho \to \min. \quad (2.3.23)$$

This problem is solved by the least-squares method.

In this section we have considered the problem of numerical simulation of wavefront correctors and described the algorithms for its solution. At this point we finish consideration of the adaptive system elements and description of the software for their numerical simulation.

We have developed a flexible and efficient numerical model for the elements of the optical feedback loop of an adaptive optics system. This model fits well with our software package describing the propagation of optical waves through the atmosphere. It has made it possible to estimate the limiting efficiency of adaptive correction and to transform the well-known mathematical model of the correction loop into real algorithms, numerical schemes, and computer programs that have passed thorough testing. In some cases, to achieve this result the corresponding mathematical equations were reformulated in forms more convenient for numerical simulation. As a result, this chapter presents and systematizes ideas on numerical simulation of the correction loop that were published in different years [66-69].

The following chapters are devoted directly to the efficiency of applying adaptive correction to the formation of beams and images in the atmosphere. The calculated results presented in these chapters are in good agreement with the known analytical estimates and experimental data. This points to the correctness of our numerical simulation methodology, the algorithms used, and the corresponding computer programs.

References

1. N.B. Baranova, B.J. Zel'dovich, "Dislocation of wavefront surface and amplitude zeros," *JETF*, **80**, p. 1789, 1981.
2. D.L. Fried, J.L. Vaughn, "Branch cuts in the phase function," *Appl. Opt.*, **31**, No. 15, pp. 2865–2882, 1992.
3. L.A. Thompson, C.S. Gardner, "Experiments on laser guide stars at Mauna Kea observatory for adaptive imaging in astronomy," *Nature (London)*, **328**, No. 16, pp. 229–231, 1987.
4. B.M. Welsh, C.S. Gardner, "Effects of turbulence-induced anisoplanatism on the imaging perfomance of adaptive-astronomical telescopes using laser guide stars," *J. Opt. Soc. Am. A.*, **8**, No. 1, pp. 69–80, 1991.
5. M.A. Vorontsov, G.W. Carhart, J.C. Ricklin, et al., "Image quality criteria for an adaptive imaging system based on statistical analysis of the speckle fields," *J. Opt. Soc. Am. A.*, **13,** No. 7, pp. 1456–1466, 1999.
6. M.C. Roggermann, B.M. Welsh, *Imaging Through Turbulence*, CRC Press, Boca Raton, FL, 1996.
7. L.C Bradley, J. Herrmann, "Phase compensation for thermal blooming," *Appl. Opt.*, **13**, No. 2, pp. 331–334, 1974.
8. J.A. Fleck, J.R. Morris, "Equivalent thin lens model for thermal blooming compensation," *Appl. Opt.*, **17**, No. 16, pp. 2575–2579, 1978.
9. J. Herrmann, "Properties of phase conjugate adaptive optics systems," *J. Opt. Soc. Am.*, **67**, No. 3, pp. 290–295, 1977.
10. V.P. Lukin, "Efficiency compensation of phase distortions of optical wave," *Sov. J. Quant. Electron.*, **4,** No. 4, pp. 923–927, 1977.
11. M.A. Vorontsov, "Method of phase conjugation for thermal blooming compensation of light fields," *Sov. J. of Quant. Electron.*, **6,** No. 10, pp. 2078- 2083, 1979.
12. H.W. Babcock, "The possibility of compensating astronomical seeing," *Publ. Astron. Soc. Pac.*, **65**, pp. 229–236, 1953.
13. V.P. Linnik, "About the principal possibility of decreasing influence of the atmosphere of a star image," *Opt. and Spectrosc.*, No. 4, pp. 401–402, 1957.

14. D. Korff, G. Dryden, R.P. Leavitt, "Isoplanicity: The translation invariance of the atmospheric Green's function," J. *Opt. Soc. Am.*, **65**, No. 11, pp. 1321–1330, 1975.
15. D.L. Fried, "Anisoplanatism in adaptive optics," *J. Opt. Soc. Am.*, **72**, No. 1, pp. 52–61, 1982.
16. E.A. Vitrichenko, V.V. Voitsekhovich, M.I. Michenko, "Estimation the field of view of adaptive telescope," Preprint No. 790, ICI RAS, Moscow, 1983.
17. E.A. Vitrichenko, M.I. Michenko, "Estimation of isoplanation radius of adaptive systems by Marrechal's creterium," Preprint No. 810, ICI RAS, Moscow, 1983.
18. R.J. Sasiela, "Strehl ratio with various types of anisoplanatism," *J. Opt. Soc. Am. A.*, **9**, No. 8, pp. 1398–1405, 1992.
19. D.L. Fried, J.F. Belsher, "Analysis of fundamental limits to artificial-guide-star adaptive-optics-system performance for astronomical imaging," *J. Opt. Soc. Am. A.*, **11**, No. 1, pp. 277–287, 1994.
20. V.P. Lukin, *Atmospheric Adaptive Optics*, Nauka, Novosibirsk, 1986. (English translation V.P. Lukin, *Atmospheric Adaptive Optics*, **PM 23**, SPIE Press, Bellingham, WA, 1996).
21. E.A.Vitrichenko, V.P. Lukin, L.A. Puchnoi, V.A. Tartakovski, *Problems of Optical Testing*, Novosibirsk, 1990.
22. "News solutions for real time wavefront sensing: adaptive optics systems, laser beams, other optical instruments," *Shack Hartmann Wavefront Sensor*, CILAS, France, May 1996.
23. Guang-ming Dai, "Modified Hartmann–Shack wavefront sensing and iterative wavefront reconstruction," *Proc. SPIE Vol.* **2201**, pp. 562–573, 1994.
24. F. Roddier, "Curvature sensing and compensation: A new concept in adaptive optics," *Appl. Opt.*, **27**, pp.1223–1225, 1988.
25. T.L. Pennigton, B.M. Welsh, M.C.Roggermann, "Performance comparison of the shearing interferometer and Hartmann wavefront sensor," *Proc. SPIE Vol.* **2201**, pp. 508–518, 1994.
26. R. Kupke, F. Roddier, D.L. Mickey, Curvature-based wavefront sensor for use on extended patterns," *Proc. SPIE Vol.* **2201**, pp. 519–527, 1994.
27. M.P. Rimmer, "Method of evaluating lateral shearing interferograms," *Appl. Opt.*, **13**, No. 3, pp. 623–629, 1974.
28. R.H. Hudgin, "Wavefront reconstruction for compensated imaging," *J. Opt. Soc. Am.*, **67**, No. 3, pp. 375–378, 1977.
29. D.L. Fried, "Least-square fitting a wavefront distortion estimate to an array of phase-difference measurements," *J. Opt. Soc. Am.*, **67**, No. 3, pp. 370–375, 1977.
30. R.J. Noll, "Phase estimates from slope-type wavefront sensors," *J. Opt. Soc. Am.*, **68**, No. 1, pp. 139–140, 1978.
31. B.R. Hunt, "Matrix formulation of the reconstruction of phase values from phase differences," *J. Opt. Soc. Am.*, **69**, No. 3, pp. 393–399, 1979.

32. K. Freischlad, C.L. Koliopoulos, "Wavefront reconstruction from noisy slope or difference data using the Discrete Fourier Transform," *Proc. SPIE Vol.* **551**, pp.74–80, 1985.
33. Takajo Hiroaki, Tohru Takahashi, "Noniterative method for obtaining the exact solution for the normal equation in least-squares phase estimation from the phase difference," *J. Opt. Soc. Am. A.*, **5**, No. 11, pp. 1818–1827, 1988.
34. Takajo Hiroaki, Tohru Takahashi, "Least-squares phase estimation from the phase difference," *J. Opt. Soc. Am. A.*, **5**, No. 3, pp. 416–425, 1988.
35. D.L Fried, "Statistics of a geometric representation of wavefront distortion," *J. Opt. Soc. Am.*, **55**, No. 11, pp. 1426–1435, 1965.
36. A.N. Bogaturov, "Solution of the set of equations for zonal reconstruction of a wavefront in adaptive optics," *Izv.V.U.Z. Fiz.*, **28**, No. 11, pp. 86–95, 1985.
37. J.W. Hardy, "Active optics: A new technology for the control of light," *Proc. IEEE*, **66**, No. 6, pp. 651–697, 1978.
38. E.A. Vitrichenko, V.V. Voitsekhovich, "Modal compensation of turbulent atmospheric distortions and isoplanarity," Preprint No. 881, ICI RAS, Moscow, 1984.
39. R.B. Shack, B.C. Platt, "Production and use of lenticular Hartmann screen," *J. Opt. Soc. Am.*, **61**, p. 1586, 1971.
40. D.S. Acton, R.C. Smithson, "Solar imaging with a segmented adaptive mirror," *Appl. Opt.*, **31**, No. 16, pp. 3161–3169, 1992.
41. V.P. Lukin, N.N. Maier, B.V. Fortes, "Point spread function calculation for adaptive telescope with Hartmann wavefront sensor," *Atm. Opt.*, **5**, No. 12, pp. 1241–1251, 1992.
42. V.I. Tatarskii, *Wave Propagation in a Turbulent Atmosphere*, Nauka, Moscow, 1967.
43. R. Cubalchini, "Modal wavefront estimation from phase derivative measurements," *J. Opt. Soc. Am.*, **69**, No. 7, pp. 972–977, 1979.
44. J. Herrmann, "Cross coupling and aliasing in modal wavefront estimation," *J. Opt. Soc. Am.*, **71**, No. 8, pp. 989–992, 1981.
45. J.Y. Wang, J.K. Markey, "Modal compensation of atmospheric turbulence phase distortion," *J. Opt. Soc. Am.*, **68**, No. 1, pp. 78–88, 1978.
46. V.P. Kandidov, I.V. Larionova, V.P. Popov, "Modal corrector of low-order phase aberrations," *Atm. Opt.*, **2**, No. 8, pp. 836–842, 1989.
47. D.L. Fried, "Statistics of a geometric representation of wavefront distortion," *J. Opt. Soc. Am.*, **55**, No. 11, pp. 1426–1435, 1965.
48. C.B. Hogge, R.R. Butts, "Frequency spectra for the geometric representation of wavefront distortions due to atmospheric turbulence," *IEEE Trans. Ant. Prop.*, **AP-24**, pp. 144–154, 1976.
49. R.J. Noll, "Zernike polynomials and atmospheric turbulence," *J. Opt. Soc. Am.*, **66**, No. 3, pp. 207–211, 1976.
50. G.C. Valley, S.M. Wandzura, "Spatial correlation of phase-expansion coefficients for propagation through atmospheric turbulence," *J. Opt. Soc. Am.*, **69**, No. 5, pp. 712–717, 1979.

51. K.V. Shishakov, V.I. Schmal'gauzen, "Polynonomial presentation of atmospheric aberrations," *Atm. Opt.*, **3**, No. 12, pp. 1244–1248, 1990.
52. J. Winocur, "Modal compensation of atmospheric turbulence induced wavefront aberrations," *Appl. Opt.*, **21**, No. 3, pp. 433–438, 1982.
53. J.H. Churnside, M.T. Tavis, H.T. Yura, "Zernike-polynomial expansion of turbulence-induced anisoplanatism," *Opt. Lett.*, **10**, pp. 258–260, 1985.
54. P.H. Hu, J. Stone, T. Stanley, "Application of Zernike polynomials to atmospheric propagation problems," *J. Opt. Soc. Am.A.*, **6**, No. 10, pp. 1595–1608, 1989.
55. V.P. Lukin, "Quazi-modal correction of image produced through random inhomogemeneity media," Sov. J. Quantum Electron., **10**, No. 5, pp. 993–1001, 1983.
56. M.A. Vorontsov, V.I. Schmal'gauzen, *Adaptive Optics Principles*, Nauka, Moscow, 1985.
57. A.V.Kudryshov, V.A.Tikhonov, V.I. Schmal'gauzen, "Dynamic characteristics of the response of a bimorph mirror," *Atm. Opt.*, **1**, No. 3, pp. 61–65, 1988.
58. S.A. Kokorowski, "Analysis of adaptive optical elements made from a piezoelectric bimorph," *J. Opt. Soc. Am.*, **69**, pp.181–187, 1979.
59. P. Jagourel, P.Y. Madec, M. Sechaud, "Adaptive optics: A bimorph mirror for wavefront correction," *Proc. SPIE Vol.* **1237**, p.394, 1990.
60. V. Samarkin, A. Alexandrov, P. Romanov, A. Kudryashov, "Novel wavefront corrector based on bimorph elements," in *Adaptive Optics for Industry and Medicine*, pp. 62-74, Starline Printing Inc., USA, 2002.
61. P.K. Mehta, "Moment actuator influence function for flat circular mirrors," *Opt. Eng.*, **29**, No. 10, pp. 1213–1222, 1990.
62. M.A. Ealey, "Deformable mirrors: design fundamentals, key performance specifications, and parametric trades," *Proc. SPIE Vol.* **1543**, pp. 36–51, 1991.
63. P. Jagourel, J.-P. Gaffard, "Adaptive optics components in laserdot," *Proc. SPIE Vol.* **1543**, pp. 76–87, 1991.
64. M.A. Ealey, "Active and adaptive optical components: The technology and future trends," *Proc. SPIE Vol.* **1543**, pp.2–34, 1991.
65. B. Hulburg, T.Barrett, L. Cuellar, D. Sandler, "High bandwidth, long stroke segmented mirror for atmospheric compensation," *Proc. SPIE Vol.* **1543**, pp.64–75, 1991.
66. V.P. Lukin, F. Yu. Kanev, P.A. Konyaev, B.V. Fortes, "Numerical model of an atmospheric adaptive optical system. Parts I, II, III," *Atm. Oceanic Opt.*, **8**, pp. 210–222, 1995.
67. V.P. Lukin, B.V. Fortes, "Efficiency of adaptive correction of images in a telescope using an artificial guide star," *OSA Tech. Digest*, **23**, pp. 92–93, 1995.

68. V.P. Lukin, B.V. Fortes, "Effect of wavefront dislocations on atmospheric adaptive optical system performance," *Proc. SPIE Vol.* **2471**, pp. 477–488, 1995.
69. B.V. Fortes and V.P. Lukin, *Adaptive Formation of Beams and Images in the Atmosphere*, Publishing House of Siberian Branch of the Russian Academy of Sciences, Novosibirsk, 1999.

CHAPTER 3

Adaptive Imaging

In this chapter we consider several problems for adaptive imaging in a turbulent atmosphere. It is well known that the angular resolution of diffraction-limited optical systems is determined by the ratio of the working wavelength λ to the diameter of the entrance pupil D [1]. When the image of an extraterrestrial object is formed, the resolution is determined by the ratio of the wavelength to the coherence length r_0 [2, 3], whose value in turn is determined by the integral structure characteristic of atmospheric turbulence C_n^2. At the places where astronomical observatories are located, the typical value of r_0 is roughly 10–20 cm for the visible spectral region.

Up-to-date astronomical instruments include a primary mirror from 2 to 10 m in size. Their actual angular atmospheric resolution, which is on the order of 1 arcsec, proves to be $D/r_0 \sim$ 10–100 times worse than the diffraction limit and is mostly independent of the aperture diameter and the wavelength. The aim of an adaptive optics system is to suppress atmospheric blurring of an image and provide for the angular resolution close to the diffraction limit. The typical structure of an adaptive telescope is shown in the schematic below.

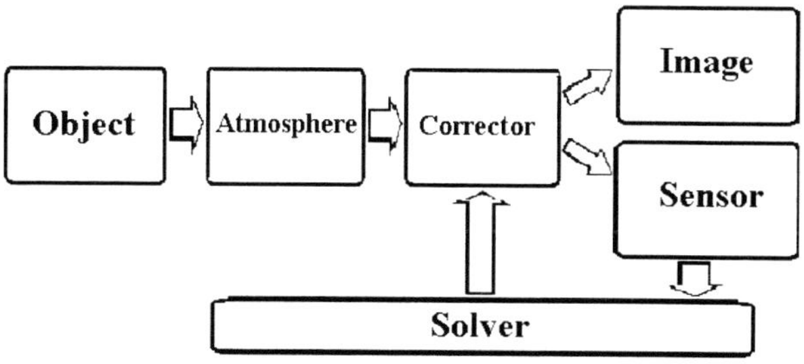

Structure of an adaptive telescope.

A wavefront distorted by turbulent inhomogeneities is reflected from the primary and secondary mirrors and arrives at the wavefront corrector. Some of the energy of the corrected radiation is directed to the image recording system and the rest goes to the wavefront sensor, which measures residual aberrations and generates the corrector control signals.

This chapter considers the aspects of telescopic imaging that are important from the viewpoint of improving the efficiency of astronomical observations; namely,

- minimization of the width of the point spread function of an atmosphere–telescope system that is due to selection of the optimal wavelength without correction of distortions;
- PSF parameters during operation of an adaptive system in the partial correction mode, i.e., at large residual aberrations; and
- efficiency of phase correction of turbulent distortions on a horizontal path at strong intensity scintillation; this problem is connected with the use of adaptive systems in optical communication.

3.1 Calculation and Minimization of Image Distortions

In this section we consider the possibility of minimizing blurring of the image of a pointlike source such as an astronomical object. The optimized parameter is the observational wavelength, and the allowance for the outer scale of turbulence is a decisive factor in the problem considered.

3.1.1 Imaging in an atmosphere–telescope system

Let the imaging system (telescope) be a thin collecting lens with the focal length f described as an amplitude-phase transparent

$$A(\vec{\rho}) = W(\vec{\rho})\exp\left(ik\frac{\rho^2}{2f}\right), \tag{3.1.1}$$

where $W(\vec{\rho})$ is the pupil amplitude function. Now study the propagation of monochromatic radiation between the plane of the entrance pupil and the system focus within the framework of the scalar theory of diffraction. Determine the slow component $E(\vec{\rho}, z)$ of the complex amplitude of the electric component of electromagnetic oscillations $\vec{E}(\vec{\rho}, z)$ and its intensity as

$$\sqrt{\frac{cn_0}{8\pi}}\vec{E}(\vec{\rho}, z) = \vec{e}E(\vec{\rho}, z)\exp(ikz - i\omega t); \tag{3.1.2}$$

$$EE^* = I, \tag{3.1.3}$$

where c is the speed of light in a vacuum, n_0 is the refractive index of the medium, $\vec{e}$ is the polarization vector, k is the wave number, and ω is the frequency of electromagnetic oscillations. In the paraxial approximation, the propagation of the complex amplitude is described by a parabolic equation of the following form:

$$2ik\frac{\partial E}{\partial z}=\left(\frac{\partial^2}{\partial x^2}+\frac{\partial^2}{\partial y^2}\right)E. \tag{3.1.4}$$

The boundary conditions of the problem are the product of the complex amplitude of the incident field E_0 and the function of the entrance pupil $A(\vec{\rho})$:

$$E(\vec{\rho},\ 0)=E_0(\vec{\rho})\times A(\vec{\rho}). \tag{3.1.5}$$

The solution of the parabolic equation (3.1.4) with the boundary condition of Eq. (3.1.5) in the focal plane $z=f$ has the form [4]

$$E(\vec{\rho},f)=E_f(\vec{\rho})=C(\rho)\iint d^2\rho' W(\vec{\rho}')E_0(\vec{\rho}')\times\exp(ik\vec{\rho}\vec{\rho}'/f);$$

$$C(\rho)=\frac{i\exp\left(-ik/\rho^2\right)}{\lambda f},$$

and can be expressed through the Fourier transform (angular energy spectrum)

$$Q(\vec{\kappa})=\iint d^2\rho' W(\vec{\rho}')E_0(\vec{\rho}')\exp(i\vec{\kappa}\vec{\rho}') \tag{3.1.6}$$

in the following way:

$$E_f(\vec{\rho})=C(\rho)Q(\vec{\rho}k/f)=c(\rho)Q[\vec{\rho}\ 2\pi/(\lambda f)]. \tag{3.1.7}$$

In the angular coordinates $\vec{\alpha}=\vec{\rho}/f$,

$$E_f(\vec{\alpha})=C(\vec{\alpha}f)\,Q(k\vec{\alpha}). \tag{3.1.8}$$

Then, the mean (long-exposure) [3] intensity distribution in the image plane can be written as follows:

$$\langle I(\vec{\alpha})\rangle=\langle E_f(\vec{\alpha})E_f^*(\vec{\alpha})\rangle=\langle|Q(k\vec{\alpha})|^2\rangle. \tag{3.1.9}$$

Consider the equation for the mean square of the modulo spectrum

$$\left\langle |Q(\vec{\kappa})|^2 \right\rangle = \iint\iint d^2\rho' d^2\rho'' W(\vec{\rho}')W(\vec{\rho}'')\left\langle E(\vec{\rho}')E^*(\vec{\rho}'')\right\rangle \exp[i\vec{\kappa}(\vec{\rho}'-\vec{\rho}'')]. \tag{3.1.10}$$

Replacing the representation of the field in terms of the complex amplitude by the representation in terms of the complex phase $\Phi(\vec{\rho})$,

$$E(\vec{\rho}) = \sqrt{I_0}\exp\{i\Phi(\vec{\rho})\}, \tag{3.1.11}$$

we derive the following equation for the expression in the angle brackets in the integrand of Eq. (3.1.10):

$$\begin{aligned}\left\langle E(\vec{\rho}')E^*(\vec{\rho}'')\right\rangle &= I_0\left\langle \exp\{i[\Phi(\vec{\rho}')-\Phi(\vec{\rho}'')]\}\right\rangle \\ &= I_0\exp\left\{-\frac{1}{2}\left\langle [\Phi(\vec{\rho}')-\Phi(\vec{\rho}'')]^2\right\rangle\right\} = I_0\exp\left[-\frac{1}{2}D_\Phi(\vec{\rho}-\vec{\rho}')\right],\end{aligned} \tag{3.1.12}$$

where D_Φ is the structure function of the complex phase and I_0 is the mean intensity of the field. After a change of variables,

$$\vec{\rho}'-\vec{\rho}'' = \vec{\rho}, \quad (\vec{\rho}'+\vec{\rho}'')/2 = \vec{r}, \tag{3.1.13}$$

Eq. (3.1.10) takes the form

$$\left\langle |Q(\vec{\kappa})|^2 \right\rangle = I_0\iint d^2\rho \exp[i\vec{\kappa}(\vec{\rho})]\exp[-\tfrac{1}{2}D_\Phi(\vec{\rho})]\iint d^2 r W\left(\vec{r}-\frac{\vec{\rho}}{2}\right)W\left(\vec{r}+\frac{\vec{\rho}}{2}\right). \tag{3.1.14}$$

For a round aperture of diameter D,

$$W(\rho) = \begin{cases} 1, & \rho \le D/2 \\ 0, & \rho > D/2 \end{cases}. \tag{3.1.15}$$

Calculation of the inner integral in Eq. (3.1.14) gives the following equation:

$$\begin{aligned}&\iint d^2 r W\left(\vec{r}-\frac{\vec{\rho}}{2}\right)W\left(\vec{r}+\frac{\vec{\rho}}{2}\right) \\ &= \left(\frac{\pi D^2}{4}\right)\frac{2}{\pi}\left[\arccos(\rho/D) - \rho/D\sqrt{1-(\rho/D)^2}\right] = \pi D^2/4\ \tau_0(\vec{\rho})\end{aligned} \tag{3.1.16}$$

and corresponds to the unnormalized diffraction-limited optical transfer function $\tau(\vec{\rho})$. Integration over the angular variable in the outer integral in Eq. (3.1.14) for the isotropic structure function leads to the following final equation for the mean intensity of the image:

$$\langle I(\vec{\alpha})\rangle = \frac{P_0}{\lambda^2}\int_0^D \rho d\rho J_0(\alpha k\rho)\tau_0(\rho)\exp\left[-\frac{1}{2}D_\Phi(\vec{\rho})\right],\quad P_0 = \frac{\pi D^2}{4}I_0. \quad (3.1.17)$$

In the neglect of turbulent amplitude fluctuations, the structure function of the complex phase transforms into the structure function of the real phase:

$$D_\Phi(\rho) \approx D_\varphi(\rho). \quad (3.1.18)$$

It is connected to the spatial spectrum of phase fluctuations F_φ by the equation

$$D_\varphi(\rho) = 4\pi^2\int_0^\infty F_\varphi(\kappa)[1 - J_0(\kappa\rho)]\kappa d\kappa. \quad (3.1.19)$$

The equations obtained [Eqs. (3.1.17) and (3.1.19)] serve then as a basis for calculating the turbulent PSF, knowledge of which is needed to solve a number of problems. The first is the minimization of the PSF width, which is considered next.

3.1.2 Minimization of the width of a turbulent PSF

Generally, the measure of the resolution of optical systems is usually the integral resolution, which is equal to the integral of the normalized optical transfer function (OTF) $\tau(\vec{\rho})$. When the images of pointlike objects are formed, for example, in astronomy, the resolution is characterized, as a rule, by the width of the PSF. Traditionally, astronomers determine the level of turbulent distortions as the PSF width, which is defined as the angular distance at which the value of the PSF halves.

Before turning our attention to the problem of adaptive correction, let us study the possibility of minimizing turbulent image distortions. By minimizing, we mean selection of the wavelength at which the PSF width achieves the minimum for the given parameters of the astroclimate: the integral intensity and the outer scale of turbulence.

The width of the turbulent PSF is determined by the largest of two values: the diffraction resolution of the instrument λ/D and the turbulent resolution λ/r_0. With a diameter much larger than the coherence length, the PSF width is determined by turbulence and at $D < r_0$, it is determined by diffraction at the aperture.

At $D < r_0$, an increase in the diameter leads to a corresponding decrease in the PSF width. When the aperture diameter reaches the coherence radius, its further increase leads only to an increase in the image intensity, because the images formed by subapertures with radii r_0 add up incoherently (in intensity).

Within the framework of the Kolmogorov theory of turbulence, wavelength variations have practically no effect on the angular resolution of a ground-based telescope because $r_0 \propto \lambda^{6/5}$ and $\lambda/r_0 \propto \lambda^{-1/6}$. As the wavelength changes from the visible (0.5 μm) to the infrared (10 μm) region, the turbulent resolution is improved only 1.65 times, whereas the diffraction resolution decreases 20 times. Therefore, a marked gain can be obtained only with a very large aperture. For the actual parameters, the PSF width usually has a slightly pronounced extreme point (minimum) in the near IR region.

The above reasons are true if the outer scale of turbulence L_0 is much larger than the aperture diameter. However, some recently published papers [5, 6] assert that the effective outer scale of turbulence on vertical paths is not as large as was thought earlier. In any case, for the apertures of current telescopes, the ratio D/L_0 does not satisfy the condition $D/L_0 \ll 1$. Therefore the angular resolution becomes a function of three parameters: aperture diameter, Fried radius, and outer scale. The normalized (to the diffraction limit) resolution becomes a function of two parameters: D/r_0 and D/L_0.

The dependence of the angular resolution on these parameters should be studied for evaluation of astroclimatic conditions in the places where astronomical observatories are situated and for the possible gain from partial adaptive correction; i.e., from equipping the telescope with an adaptive system having low spatial resolution. First of all, this is important for designing the "simplest" adaptive systems that correct the total wavefront tilt, because the variance of its fluctuations depends significantly on the outer scale.

In our calculations, Eqs. (3.1.17) and (3.1.19) were used and the spatial spectrum of phase fluctuations was set in the form of the Karman spectrum [7, 8]:

$$F_\varphi(\kappa) = 0.489 r_0^{-5/3} \left(\kappa^2 + L_0^{-2}\right)^{-11/6}. \quad (3.1.20)$$

The structure function corresponding to this spectrum and calculated by numerical integration of Eq. (3.1.17) is shown in Fig. 3.1.1.

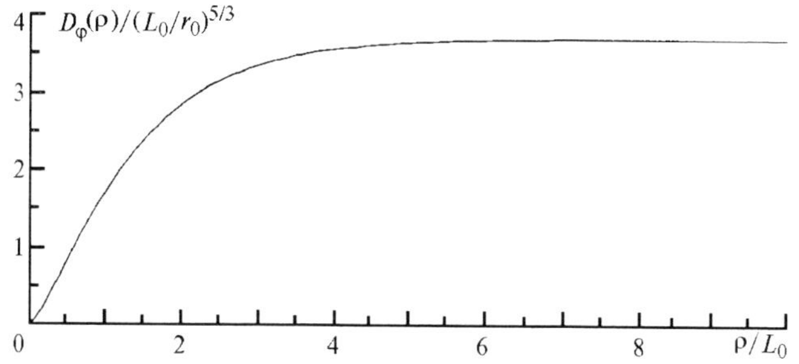

Figure 3.1.1. Phase structure function corresponding to the spectrum of Eq. (3.1.20) and calculated by numerical integration of Eq. (3.1.17).

As can be seen from the figure, the structure function saturates at $3.7(L_0/r_0)^{5/3}$, reaching the level of 0.9 saturation at $\rho/L_0 = 3$. This agrees with the theoretical value for the phase variance:

$$\sigma^2 = 0.78k^2 C_n^2 L L_0^{5/3} = 1.85(L_0 / r_0)^{5/3} .$$

Consider now how the PSF changes as the outer scale L_0 decreases for the fixed ratio D/r_0. The results calculated for the PSF [as a function of $\alpha/(\lambda/D)$, where α is the angle] are shown in Fig. 3.1.2. The calculation has been made for $D/r_0 = 10$. The ratio D/L_0 took the values of 1, 10^{-1}, and 10^{-2}. It can be seen that the PSF width depends markedly on the outer scale. The axial value of the PSF [Strehl ratio, SR] almost doubles as the outer scale decreases from 100 to 1 aperture diameter.

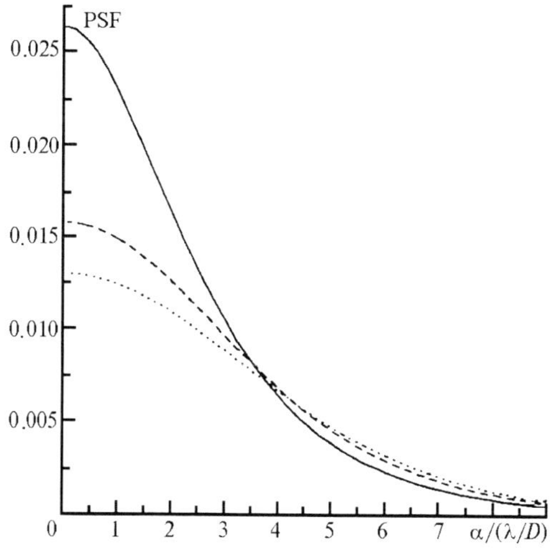

Figure 3.1.2. Turbulent PSF for different values of the outer scale: $D/L_0 = 1$ (solid curve), 10^{-1} (dotted curve), and 10^{-2} (dashed curve). The normalized aperture diameter $D/r_0 = 10$.

Let us study in detail the dependence of the PSF width and the SR on L_0/D at different values of D/r_0 (Fig. 3.1.3). It can be seen that the outer scale affects most strongly the PSF parameters at the aperture diameter $D/r_0 \approx 10$. At small D/r_0, turbulence has a weak effect on the PSF and therefore the effect of its outer scale is also weak.

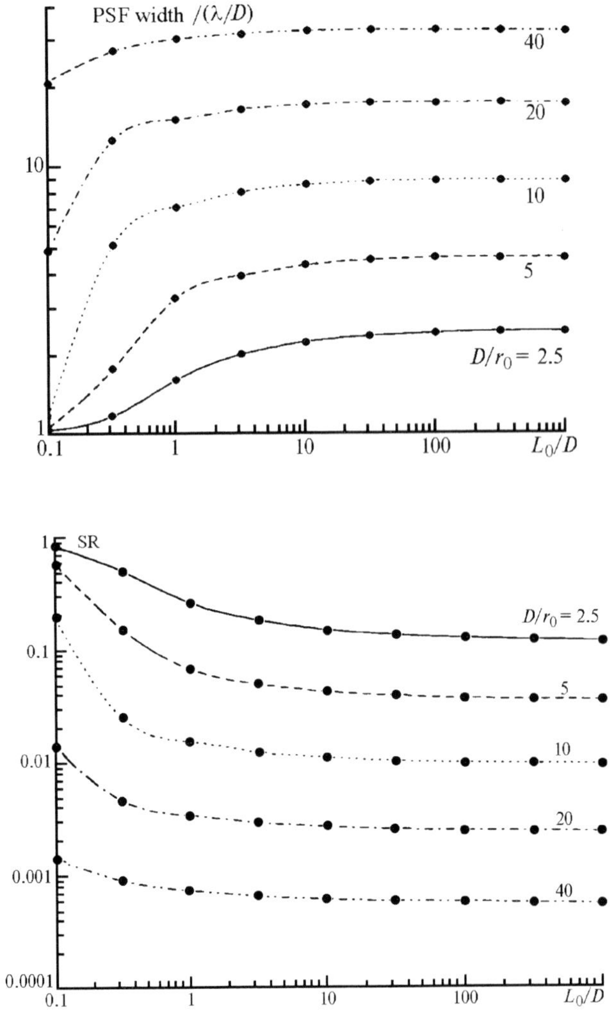

Figure 3.1.3. Dependence of PSF width and SR on L_0/D at different values of D/r_0.

For large values of D/r_0 ($D/r_0 > 20$), the effect of the outer scale also weakens, but for another reason. From a mathematical point of view, this can be explained as follows: In the integral [Eq. (3.1.17)], the domain significant for integration is determined by the exponent of the phase structure function $D_\varphi(\rho)$. At $\rho > r_0$, the value of the exponential factor becomes much less than unity. Consequently, the domain of significant integration is limited by the inequality $\rho < r_0$. Therefore, as D/r_0 increases, the value of L_0 at which the structure function deviates markedly from the power dependence decreases.

Let us study now the wavelength dependence of the PSF width for the finite outer scale. Consider a particular example. For places with a good astroclimate, the typical value of the parameter r_0 in the visible region (0.5 μm) is about 20 cm. The corresponding integral structure characteristic of refractive index fluctuations is $C_n^2 L = 2.2\times10^{-13}$, and for an arbitrary wavelength it is $r_0 = 20$ cm $\times(\lambda/0.5\ \mu\text{m})^{6/5}$.

Figure 3.1.4 shows the wavelength dependence of the PSF width for the outer scale of turbulence from 0.1 to 100 m. The calculations have been made for telescopes 1–8 m in diameter.

At $L_0 > 10$ m, an increase in the wavelength leads to some decrease in the PSF width. However, this gain is noticeable only for large apertures. For $D = 8$ m, the transition from $\lambda = 0.5$ to $\lambda = 4.5$ μm causes a decrease in turbulent blurring from 0.5 arcsec to 0.25 arcsec at $L_0 = 1$ m. For apertures of 1–2 m, the transition into the infrared region gives rise to diffraction effects and an increase in the PSF width. At the outer scale $L_0 = (0.1–0.3)$ m, the PSF width is determined by the diffraction effects in almost the entire region of the wavelength; i.e., it increases linearly with growing λ.

A quantitative analysis shows that at $L_0 = 1$ m the wavelength dependence of the telescope's resolution is most pronounced. Thus, the optimal wavelength increases from 1 μm for a 1-m aperture to 2 μm for a telescope 8 m in diameter. As this takes place, the minimum PSF width changes respectively from 0.35 to 0.05 arcsec. For an 8-m telescope, this gives an almost eightfold gain compared with the visible region (0.4 arcsec at 0.5 μm). The gain in resolution achieved at the optimal wavelength increases almost linearly as the telescope diameter increases and roughly equals D/L_0.

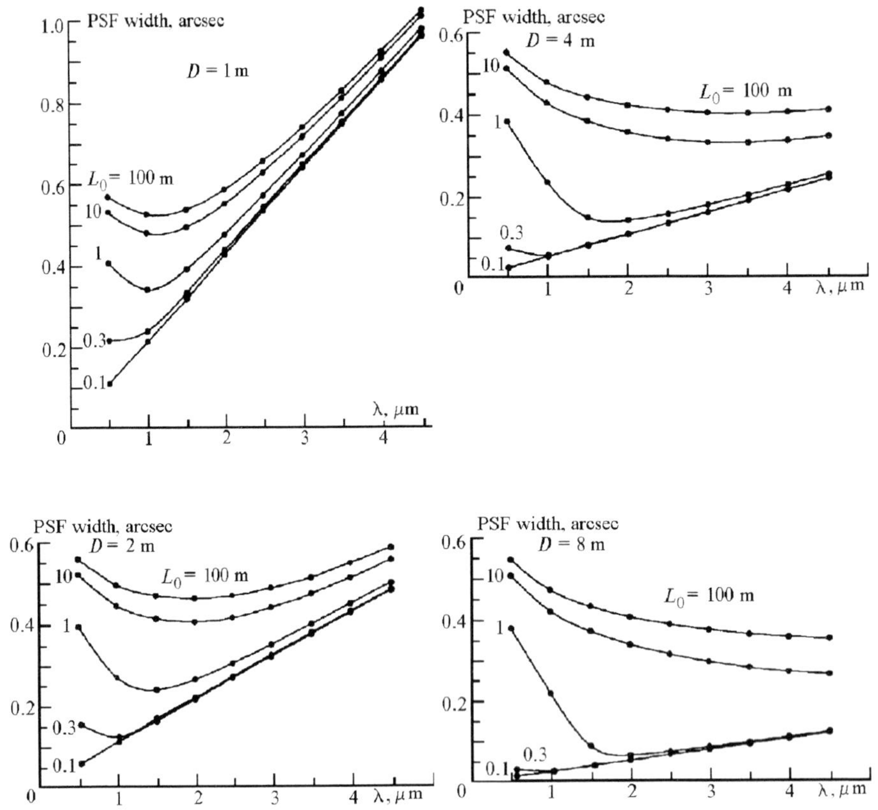

Figure 3.1.4. Wavelength dependence of the PSF width for the outer scale of turbulence from 0.1 to 100 m and different diameters D of the telescope.

3.2 Study of Angular Resolution and Contrast in Large Residual Wavefront Distortions

Up-to-date telescopes have apertures whose diameter is 10 times larger than the atmospheric coherence length for the visible region. To correct turbulent distortions completely, the adaptive system in such telescopes should have hundreds of correction channels. The complexity and cost of adaptive optics devices grows rapidly as the requirements for spatial resolution increase.

Many projects aimed at developing adaptive systems for large telescopes involve the development of technological solutions for a relatively small number of degrees of freedom in the corrector. In the infrared region, such systems can provide almost complete compensation for turbulent distortions, whereas in the visible region the compensation is only partial.

A complete correction usually describes a correction providing for an SR higher than 0.5. A partial correction may also give a considerable gain. For example, if the SR is 0.01 without correction and 0.1 after correction, then the time of image recording can be decreased by an order of magnitude.

In this section we consider the results of numerical simulation of an adaptive system operating in the mode of partial correction. We are interested first in the dependence of the PSF width on the level of residual distortions. Consider first the residual distortions that are due to finite spatial resolution of the wavefront corrector.

3.2.1 Effect of a corrector's spatial resolution on PSF parameters

Let us begin our discussion with a modal corrector capable of correcting N Zernike polynomials. The theory of the modal corrector as applied to compensation for turbulent distortions has been the subject of several theoretical papers [9-12]. The efficiency of the modal corrector can be estimated from the results obtained in Ref. [10], where the variance of residual phase distortions of the corrected wave was calculated. This variance is a function of the normalized diameter of the telescope aperture D/r_0 and the number of Zernike polynomials compensated by the modal corrector. The corresponding equation has the form [10]:

$$\sigma_N^2 = \frac{1}{S}\iint_A d^2\rho\langle\Delta\varphi^2\rangle = C_N\,(D/r_0)^{5/3},\quad \Delta\varphi = \varphi_t(\vec{\rho}) - \sum_1^N a_l Z_l\,(2\vec{\rho}/D),\quad (3.2.1)$$

where S is the area of the aperture A, φ_t are turbulence-induced phase distortions, Z_l are Zernike polynomials, and a_l are the coefficients determined from the condition of the minimum variance σ_N^2. The values of the coefficients C_N are given in Table 3.2.1, where $N = 1$ corresponds to compensation for the constant component, $N = 3$ to compensation for the constant component and tilts (linear aberrations), $N = 6$ to compensation for aberrations up to and including square ones, and so on.

Table 3.2.1. Values of coefficients of C_N.

N	1	3	6	10	15	21
C_N	1.03	0.134	0.065	0.040	0.028	0.021

For large values of N there exists the approximate equation [10]

$$C_N \approx 0.2944 N^{-\sqrt{3}/2} \qquad (3.2.2)$$

for estimating the degree of residual distortions for a given parameter N of the modal corrector and vice versa, the number of polynomials to be compensated from the given level of residual distortions.

Let us estimate the number of polynomials to be compensated for the complete correction to be achieved. Take the condition $\sigma_N^2 = 1$ as a quantitative criterion. In terms of wavefront aberrations or an ikonal, this corresponds to the $\lambda/6$ criterion for the root-mean-square value of wavefront aberrations. At such a level of distortion, the Strehl ratio is

$$\mathrm{SR} = \exp\left(-\sigma_N^2\right) = 1/e \approx 0.37. \tag{3.2.3}$$

Substituting the condition $\sigma_N^2 = 1$ into Eq. (3.2.2), we obtain the estimate of the needed degrees of freedom of the modal corrector:

$$N = 0.244\,(D/r_0)^{1.92}. \tag{3.2.4}$$

The dependence of N on the normalized aperture diameter is given in Table 3.2.2.

Table 3.2.2. Dependence of N on normalized diameter.

D/r_0	10	20	30	40	50
N	20	78	170	295	454

Compare these theoretical estimates with the results of our numerical experiment conducted in the following way: Turbulent distortions were simulated in the approximation of a phase screen with the given parameter r_0. Intensity scintillation was ignored. Wavefront distortions were assumed to be known (the ideal sensor model was used). The definition of corrector control was based on minimization of the integral square correction error.

The calculations have been performed for the normalized aperture diameter $D/r_0 = 10$–30. The second parameter of the problem was the number N of corrected polynomials, which varied in the range from 3 to 28 corresponding to correction of aberrations with radial polynomials from the first to the fifth degree. The computational grid was 128×128, and averaging was made over 100 realizations of random phase screens.

Figure 3.2.1 shows the PSF for $D/r_0 = 20$. According to Table 3.2.2, for complete correction of distortions at such an aperture diameter, a modal corrector with $N = 78$ is needed. However, we see that even at $N = 10$, the first diffraction fringes are observable in the central part of the PSF. In this case, the intensity of the PSF central lobe is 66 times lower than in the diffraction-limited system (SR = 0.015 = 1/66).

It can be seen from Fig. 3.2.1 that the partially corrected PSF consists of two components, one having a width roughly equal to the width of the turbulent PSF, and another having a diffraction-limited angular size equal to λ/D.

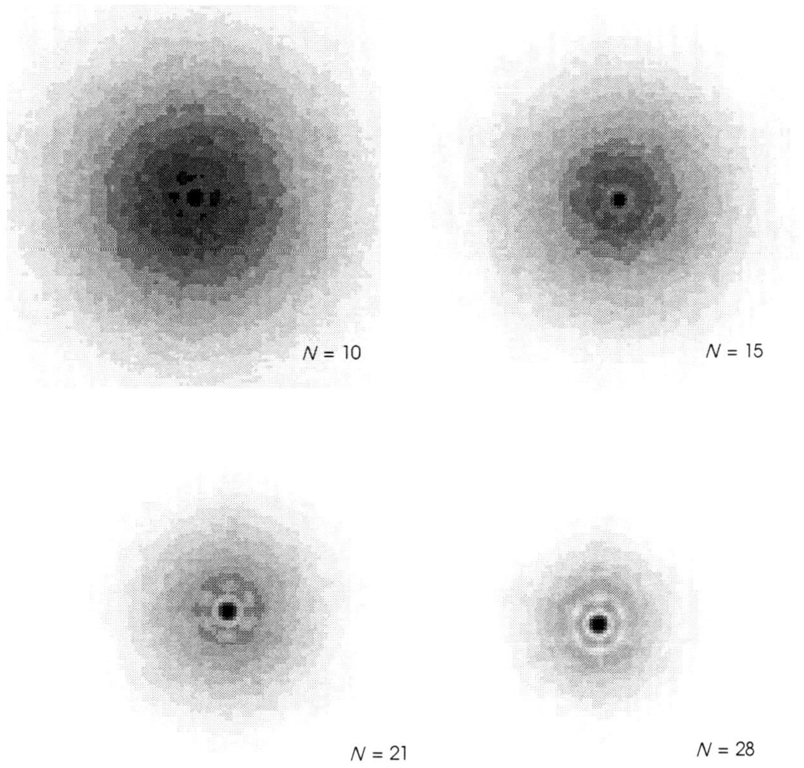

Figure 3.2.1. Two-dimensional distribution of the PSF intensity at modal correction; the normalized aperture diameter $D/r_0 = 20$; N is the number of corrected modes (Zernike polynomials).

Consider the radial cross sections of the PSF normalized to an on-axis value that are shown in Fig. 3.2.2. This normalization makes it easier to see the PSF structure at different levels of residual distortions. The value of the Strehl ratio is shown in Fig. 3.2.2 next to the number of corrected aberrations. At $D/r_0 = 10$ and $N = 21$ it is equal to 0.44. It follows from Table 3.2.2 that at $D/r_0 = 10$ the number of polynomials to be corrected for achieving SR = 0.37 is $N = 20$. The results of numerical simulation agree well with analytical estimates.

It is interesting to study how the PSF contrast parameter changes as the normalized aperture diameter and the number of corrected aberrations increase. Let us define the contrast as the ratio of the on-axis intensity to the intensity at the level of the first light fringe.

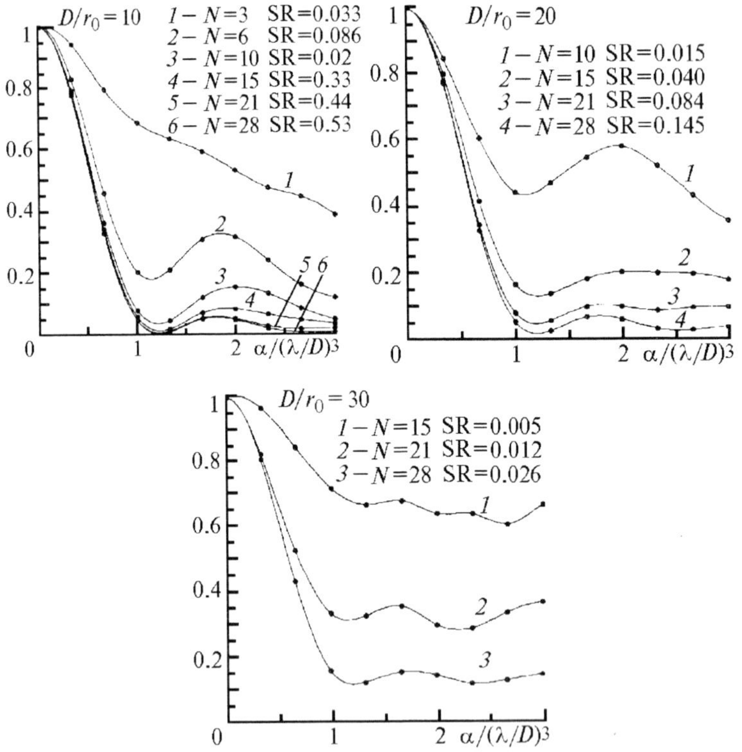

Figure 3.2.2. The PSF with the use of a modal corrector. Parameter *N* corresponds to the number of Zernike polynomials. The PSF is normalized to the on-axis value.

In a problem of resolving two pointlike objects spaced by the angular distance λ/D, the contrast determines the limiting brightness ratio at which the dim object is still seen against the background of the brighter one (Fig. 3.2.3).

It can be seen from Fig. 3.2.3 that at the same SR, a far higher contrast is achieved at larger aperture diameters. Consequently, a pronounced diffraction nucleus against the background of a turbulent spot can be seen earlier (for large residual distortions) at a higher level of initial turbulent aberrations. This is explained by the fact that at the low intensity level of the turbulent component of the partially corrected PSF, its diffraction-limited segment is characterized by a higher contrast at the same SR.

For example, if for the uncorrected image the SR equals 0.001, then a partial correction increasing the on-axis intensity up to 0.01 leads to the diffraction component of the PSF with a contrast parameter of about 10. To obtain the same contrast at an initial level of distortions corresponding to SR = 0.01, turbulent distortions should be corrected down to the level corresponding to SR = 0.1. This is achieved at the far smaller value of the variance of initial phase distortions. In parallel with calculation of the PSF according to Eq. (3.2.1), the sample-average variance of residual phase distortions was calculated. Figure 3.2.4 shows the dependence of the SR on this variance.

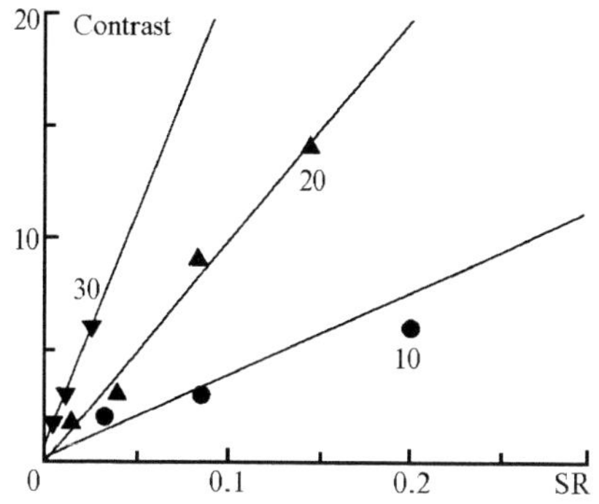

Figure 3.2.3. Dependence of PSF contrast on the Strehl ratio at modal correction for the aperture diameter $D/r_0 = 10$, 20, and 30. The straight-line slope coefficients obtained from fitting by linear regression are equal to 37, 99, and 206, respectively. The contrast is defined as the ratio of the PSF intensity on the optical axis ($\alpha = 0$) to the intensity at the level of the first light fringe ($\alpha \cong 2\lambda/D$).

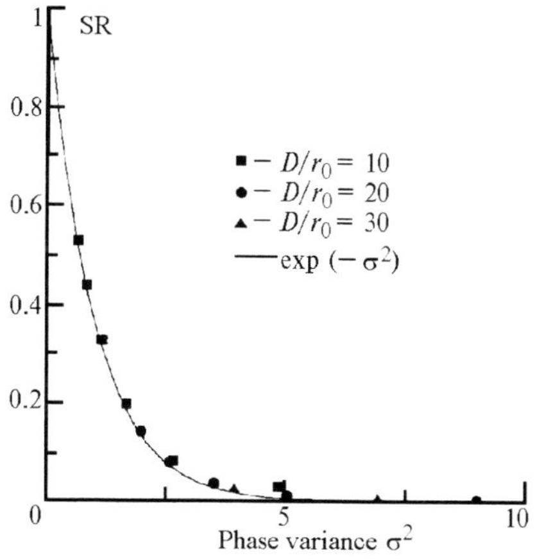

Figure 3.2.4. Dependence of SR on variance of residual phase distortions.

Thus, a sufficiently contrasting diffraction nucleus of the PSF can remain up to an SR on the order of 0.01 and for $D/r_0 > 30$, probably to even lower values. This fact, generally speaking, is not obvious. Actually, it could be expected that the tenfold decrease of the on-axis intensity of the image would lead to its almost threefold widening because the intensity is inversely proportional to the square effective size at the unchanged shape of the intensity distribution in the image plane.

However, in compensating for the lowest aberrations, the spatial spectrum of wavefront distortions changes significantly. Uncorrected small-scale aberrations lead to redistribution of the corrected PSF energy into far wings (compared with the turbulent PSF at the same variance of phase distortions). In this case, the PSF width differs only slightly from the diffraction-limited one. This allows the astronomic measurements of angular positions of objects to be conducted with an accuracy close to the diffraction limit at even relatively weak (in terms of SR) correction.

Nearly diffraction-limited resolution at low SR values is also possible with the use of other classes of wavefront correctors. Let us consider deformable mirrors of the zonal type and segmented correctors from this point of view. For both the zonal-type mirrors and segmented correctors, the equation for the variance of the residual error of phase correction has the following form:

$$\sigma^2 = C(d/r_0)^{5/3}, \qquad (3.2.5)$$

where d is the characteristic scale of the corrector, i.e., the distance between control points of a zonal mirror, or the size of an element of the segmented mirror, and the coefficient C depends on corrector's features. For approximate estimation, we can take $C = 0.2$ for the corrector with the Gaussian response function [13] and $C = 0.134$ for the segmented mirror, the tilt and position of whose elements are corrected. In the latter case, we have used the theory of the modal corrector assuming that every segment eliminates the piston mode and wavefront tilts within its subaperture.

The ratio of the aperture diameter to the size of the corrector zone at $\sigma^2 = 1$ is determined as

$$N = \frac{D}{d} = C^{3/5} D/r_0 \,. \qquad (3.2.6)$$

Table 3.2.3 gives the values of N rounded to the closest integer for correctors of both types. The upper row gives the values of the normalized aperture diameter (10–50); the medium row corresponds to the segmented corrector, and the lower one is for the deformable mirror.

Table 3.2.3. Rounded values of *N*.

D/r_0	10	20	30	40	50
$C = 0.134$	3	6	9	12	15
$C = 0.200$	4	8	11	15	19

Consider now the results of numerical simulation. Figure 3.2.5 shows the results calculated for a segmented corrector with cellular segment packing at $D/d = 11$. This spatial resolution is sufficient for complete correction at $D/r_0 \sim 30$. It can be seen from the figure that at $D/r_0 = 50$ the normalized PSF is almost indistinguishable from the diffraction-limited one within the first light fringe, whereas the SR is roughly equal to 0.1, i.e., the on-axis intensity is an order of magnitude lower than the diffraction-limited value.

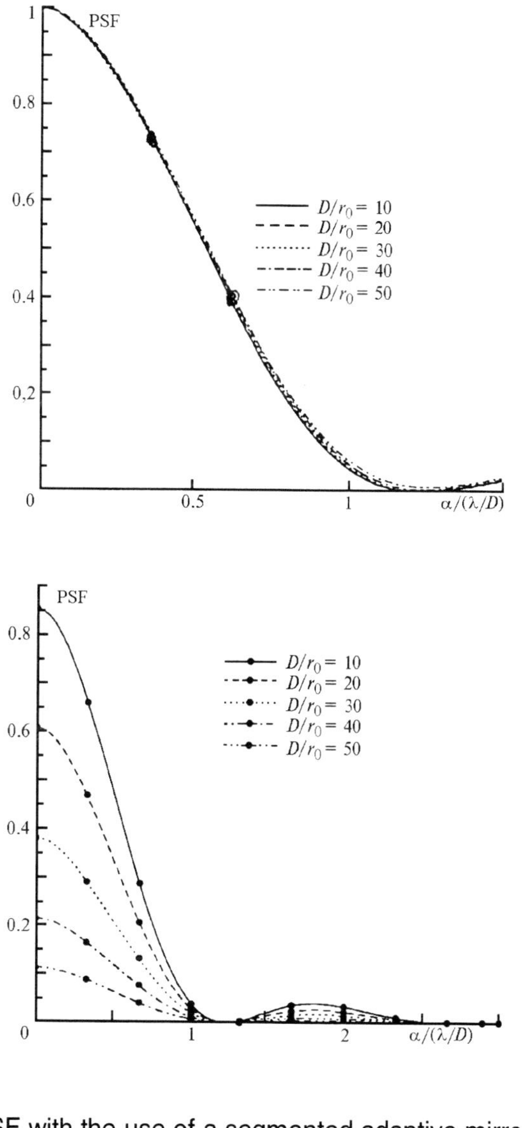

Figure 3.2.5. A PSF with the use of a segmented adaptive mirror consisting of 84 hexagonal segments, each controlled in both position and tilt: Upper panel, PSF normalized to the on-axis value; lower panel, PSF normalized to the diffraction maximum.

Similar calculations have been made for a deformable mirror with a Gaussian response function. The results are shown in Fig. 3.2.6. According to Table 3.2.3, the resolution $N = 9$ is sufficient for complete correction only at the diameter $D/r_0 = 20$; however, even at $D/r_0 = 50$, the PSF has a diffraction nucleus with a contrast of about 6.

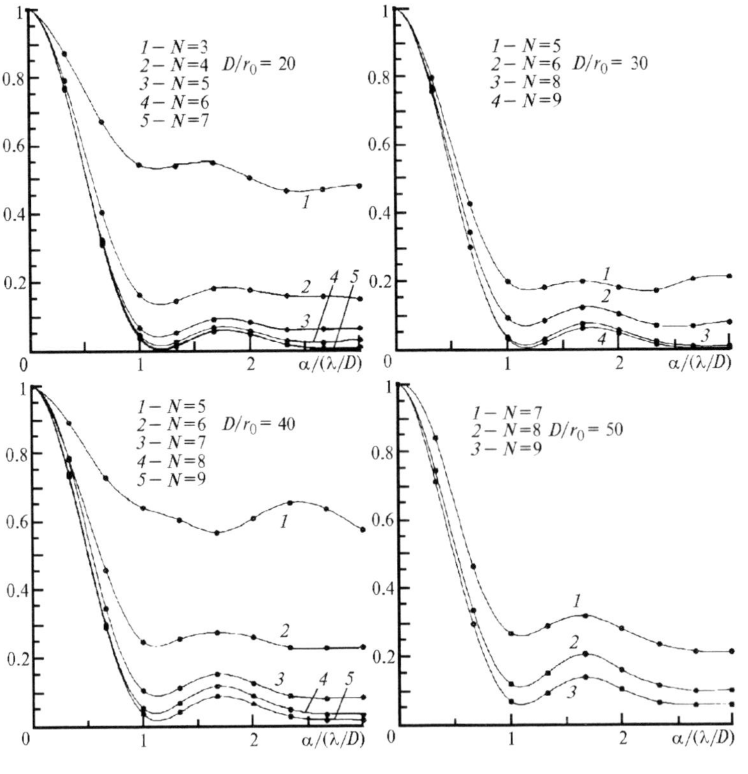

Figure 3.2.6. A PSF with the use of a deformable adaptive mirror with a Gaussian response function. The parameter N corresponds to the number of control points on the aperture diameter; the PSF is normalized to the on-axis value.

3.2.2 Measurement of distortions under limited photon flux

Let us consider here the Shack–Hartmann sensor, which is used most widely in astronomical applications. In this sensor, the spatial resolution is determined by the subaperture size and the temporal resolution is determined by the time of signal accumulation. The product of the subaperture area, accumulation time, and the intensity of reference radiation determine the number of photoelectrons

N at the sensor output. This number is the parameter determining the signal-to-noise ratio.

Let us now study the effect of limited photon flux on the quality of correction using as an example a numerical experiment for a Shack–Hartmann sensor with a 10×10 subaperture array [14]. To reconstruct the pattern of aberrations, we use the modal reconstruction algorithm and calculate coefficients of 28 Zernike polynomials. Let the ideal corrector reconstruct exactly all aberrations and let the diameter of the telescope aperture be $D = 10r_0$. Thus, the diameter of the sensor subaperture is equal to the coherence length. The correction delay is assumed to be negligibly short. Turbulent distortions are modeled in the approximation of the phase screen; intensity scintillations are ignored.

Figure 3.2.7 shows the PSF obtained in the numerical experiment. The mean number of photons N_{ph} falling on one sensor subaperture for one measurement cycle varied from 3 to 100. The figure shows the PSF for $N_{ph} = 5$, 10, and 100. At $N_{ph} = 100$, the SR equals 0.38. This value is close to the result obtained for the modal corrector with the same number of aberrations to be corrected (28) with the use of the ideal sensor model (SR = 0.53). The difference is explained by the finite resolution of the sensor and a small error due to noise. At $N_{ph} = 5$, the on-axis value of the PSF almost halves and the SR equals 0.19. A comparison of PSFs normalized to the on-axis value shows that the PSF width changes only slightly and the contrast equals 10 even at this noise level.

A further decrease in the intensity of the reference wave leads to a rapid increase in the variance of residual phase distortions σ^2 and a decrease in the SR. Figure 3.2.8 shows the dependencies of these parameters on the number of photons. At $N_{ph} = 3$, the variance σ^2 reaches 9, and the SR value drops to 0.06. This value is almost twice as large as the value resulting from tilt correction (0.0033).

Thus, in this case, the increase in residual distortions with the decreasing brightness of the reference source again leads to manifestation of the two-component structure of the partially corrected PSF: a diffraction lobe and turbulent spot.

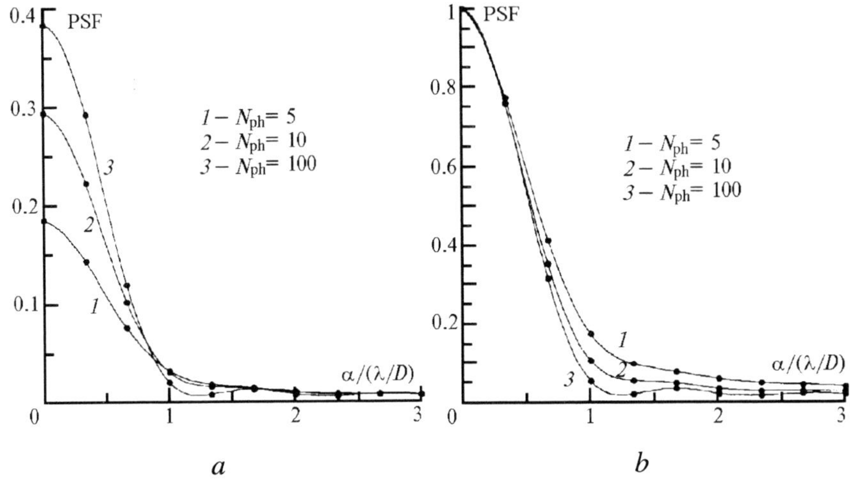

Figure 3.2.7. Modeled results for an adaptive telescope with a Shack–Hartmann sensor. The normalized diameter of the telescope aperture $D/r_0 = 10$. The sensor includes a 10×10 lenslet array. Wavefront aberrations were estimated using the modal algorithm (28 Zernike polynomials). The parameter N is the mean number of photons on a subaperture for one exposure. The PSF is normalized to the diffraction maximum (a) and to the on-axis value (b).

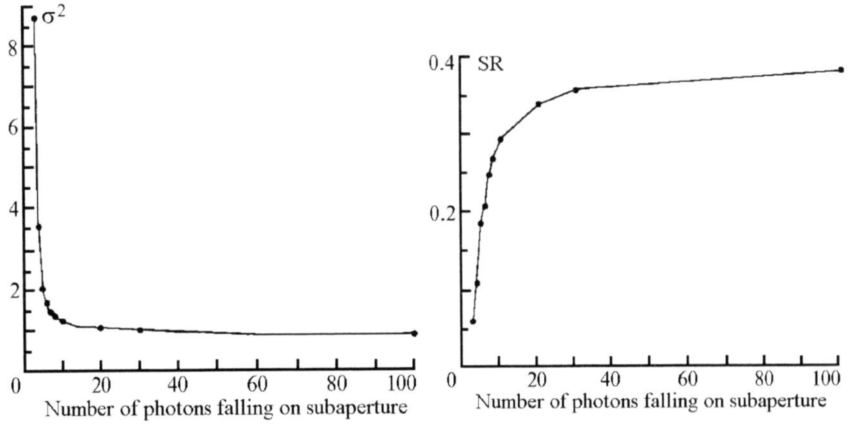

Figure 3.2.8. Variance of the residual error of phase correction σ^2 and SR vs. brightness of the reference source. The parameters of the numerical experiment are the same as in Fig. 3.2.7.

3.2.3 Effect of cone anisoplanatism

The quest to obtain a measurement accuracy that is not dependent on the presence of a sufficiently bright star in the field of view of an adaptive system has stimulated the development of artificial laser reference sources. These are formed by the atmospheric backscattering of laser pulses. The backscattered radiation is distorted by the atmosphere as a spherical wave, while the radiation of an astronomical object is a plane wave. This gives rise to a new type of correction error called the "error of cone anisoplanatism."

Consider how this error affects the PSF width [15]. We assume that the adaptive system has infinite spatial and temporal resolution, noise is absent, and the sole source of error is cone anisoplanatism. The atmosphere is modeled by a set of random phase screens. Propagation of the wave through phase screens is calculated by a ray-optics approximation that ignores the diffraction effects and intensity scintillation. It is assumed that fluctuations of the incoming angle are corrected against a bright star.

For this study, we took an altitude profile of the turbulence intensity corresponding to Gurvich's model of "best conditions" [16]. The aperture diameter was 10 m; this value corresponds to the diameters of the twin Keck telescopes. The height of the artificial beacon H was taken to be equal to 10 and 100 km, corresponding to the typical schemes based on Rayleigh scattering and resonance scattering in the sodium layer. The wavelength λ was varied from 1 to 5 μm for the Rayleigh beacon ($H = 10$ km) and from 0.5 to 2 μm for the sodium one ($H = 100$ km).

Figure 3.2.9 shows the PSFs for these two cases. It can be seen that the use of the Rayleigh reference source is efficient only in the IR region ($\lambda > 1$ μm). The use of the sodium beacon allows SR > 0.7 and the diffraction-limited angular resolution to be achieved in the infrared region ($\lambda > 1$ μm). In the visible region, the PSF width is also close to the diffraction limit at SR = 0.23 and the contrast is higher than 10. With allowance for the fact that the angular coordinate is normalized to the ratio λ/D, the angular resolution in the visible region is higher than in the infrared.

Summarizing the analysis of the partial correction of turbulent distortions, we would like to note the following: Three versions of idealized adaptive systems have been considered. In each case, only one factor determining residual distortions of the PSF was taken into consideration, namely, spatial resolution of the corrector in the first case, sensor noise in the second case, and cone anisoplanatism in the last case. In all cases, it turned out that the PSF width increases slowly and can remain nearly diffraction limited at rather high residual distortions [17].

An analogy can be drawn between this result and the PSF calculations for the spectrum with the finite outer scale that were presented in the Sec. 3.1.2. Partial correction is, first of all, the correction of the lowest aberrations corresponding to large-scale inhomogeneities of the turbulent atmosphere. This is especially true when the finite spatial resolution of the correction device plays a major role. Aberrations caused by inhomogeneities larger than the size of the corrector element are compensated for, and the spatial spectrum of residual distortions is similar to the spectrum with the finite outer scale equal to the size of the corrector element.

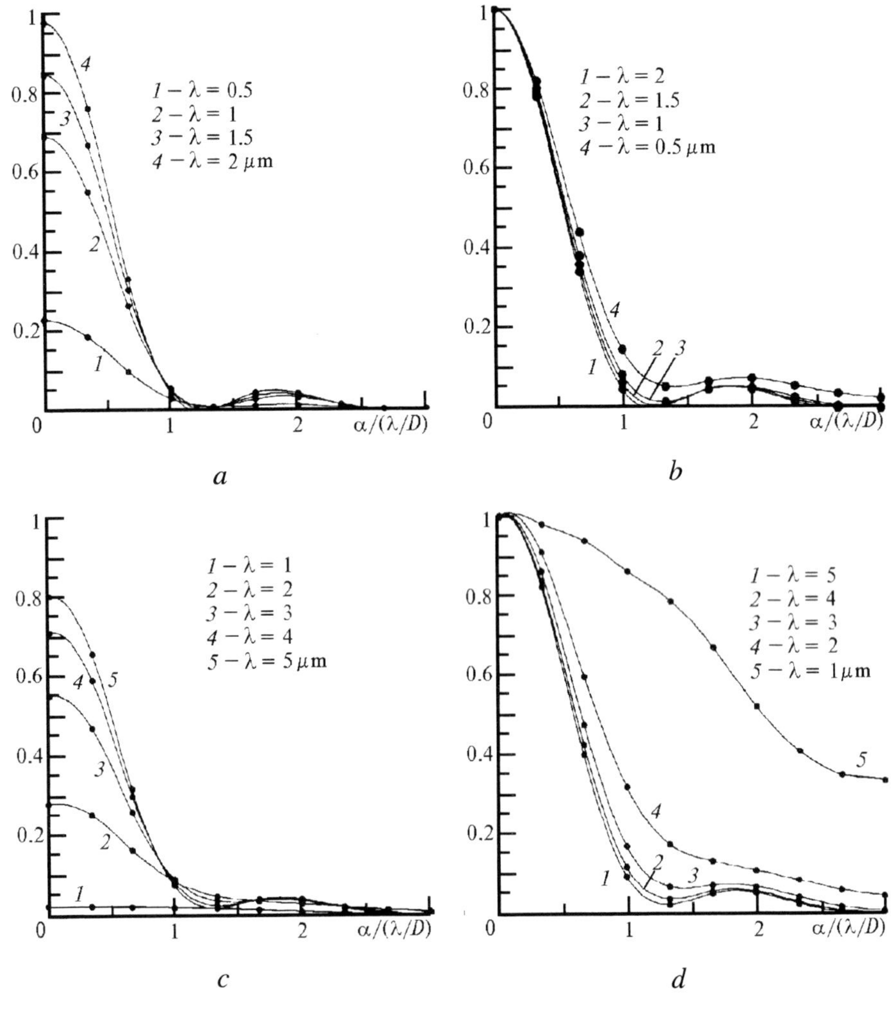

Figure 3.2.9. PSF for different wavelengths with adaptive correction against a laser guide star: (a and b) sodium laser guide star, $H = 100$ km; (c and d) Rayleigh laser guide star, $H = 10$ km; (a and c) PSF normalized to the diffraction maximum and (b and d) to the on-axis value.

The appearance of the diffraction nucleus corresponding to the central lobe of the diffraction-limited PSF becomes clear from this. The explanation can be drawn from consideration of the chain structure function, OTF, and PSF. The phase structure function $D(r)$ reaches the saturation level of $2\sigma_\varphi^2$ at a spacing on the order of the outer scale of turbulence L_0 or the size of the corrector element

d. The optical transfer function of the atmosphere $\tau(\rho) = \exp[-D(\rho)/2]$ decreases to the level $\exp(-\sigma_\varphi^2)$ and also ceases to change. Thus, the OTF consists of a narrow lobe with the thickness of L_0 or d and a wide pedestal whose level is $\exp(-\sigma_\varphi^2)$ times lower than the central maximum. The OTF can be thought of as consisting of two components: $\mathrm{OTF} = \mathrm{OTF}_1 + \mathrm{OTF}_2$. The PSF is connected with the OTF through the two-dimensional Fourier transform F; i.e., $\mathrm{PSF} = F(\mathrm{OTF}_1 + \mathrm{OTF}_2) = \mathrm{PSF}_1 + \mathrm{PSF}_2$. Since the width of the Fourier transform is inversely proportional to the width of the initial function, the narrow component OTF_1 transforms into the wide component PSF_1 (turbulent spot), whereas the wide component OTF_2 transforms into the narrow diffraction lobe PSF_2.

3.3 Phase Correction of Turbulent Distortions Under Strong Intensity Scintillation

In Sec. 3.2, we discussed the characteristics of phase correction with a large variance in residual phase fluctuations. In this section we consider another limiting case—strong intensity scintillation—assuming that the adaptive system has infinite spatial and temporal resolution in the correction of phase distortions.

It is well known that the phase distortions a wave acquires while passing through an optically inhomogeneous medium transform into modulation of the spatial intensity distribution during further propagation of the wave. At sufficiently deep modulation, points with zero intensity may arise. If the wave is described in terms of the complex amplitude U, then these points arise at the places of line intersection (or contact) where its real and imaginary parts are zero. If ReU and ImU alternate their sign from positive to negative when going through these lines, then such intersection points are the points of wavefront dislocations [18]. From the viewpoint of adaptive phase correction, it is important that the two-dimensional phase distribution becomes discontinuous when dislocations arise [19].

With such discontinuities, the error of approximation of the wavefront by an adaptive mirror increases markedly. Application of specialized correctors has no effect in the general case because the dislocations arise at randomly located points when correcting for turbulent distortions.

The algorithms for mapping aberrations of the reference wave that are currently used in most wavefront sensors output the continuous function of the transverse coordinates. In actuality, they filter the vortex of the measurement vector [20].

Consider the results of numerical experiments that demonstrate the effect of intensity scintillation and dislocations on the correction of turbulent distortions. Two aspects are of practical and purely scientific interest: how strongly the loss of the amplitude information affects the efficiency of phase correction, and how strongly the loss of the information contained in the vortex part of phase measurements decreases the efficiency of adaptive correction.

Numerical experiments have been conducted for two correction schemes [21, 22], which are shown in Fig. 3.3.1. The upper panel shows compensation for distortions. A plane wave passes through the layer of a turbulent medium of length L and arrives at the receiving aperture of the adaptive system having a diameter D. The corrected wave is focused by the collecting lens onto a point photodetector recording the intensity at the system's focus. The criterion of the correction quality is the ratio of this intensity to its diffraction-limited value averaged over an ensemble, i.e., the Strehl ratio.

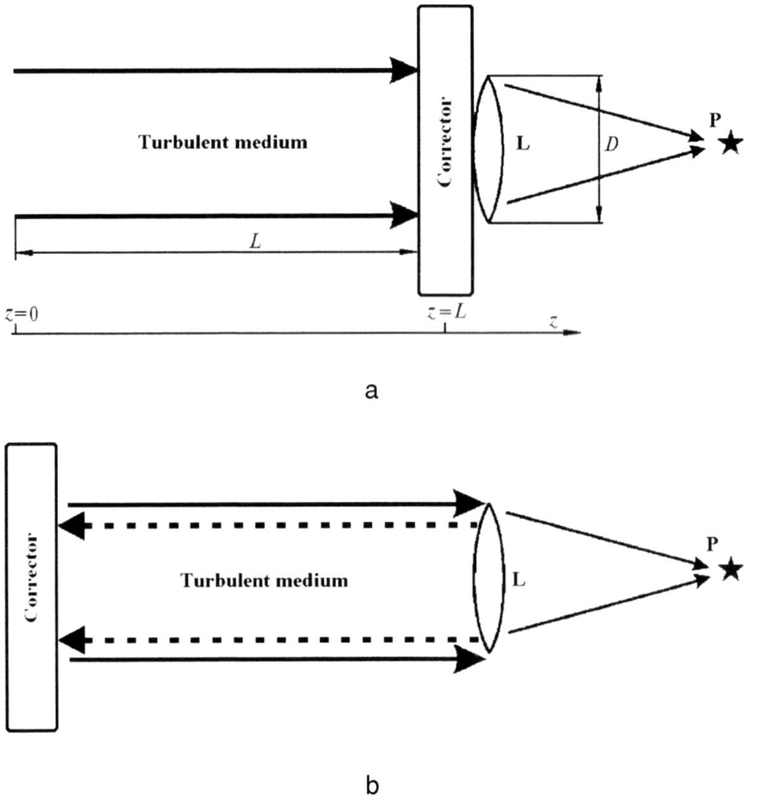

Figure 3.3.1. Optical layout of the numerical experiment: (a) compensation scheme for phase distortions and (b) phase conjugation scheme.

The lower panel shows the phase conjugation scheme. The recording system consisting of the lens L and the photodetector P is still in the plane $z = L$, but the adaptive system is in the plane $z = 0$. A plane reference wave propagates from the plane $z = L$ to the plane $z = 0$. In the plane $z = 0$, the adaptive system measures distortions of the reference radiation and corrects the wave according to the PC algorithm.

For each of the two schemes, two versions of the phase measurement algorithms have been considered. In the first version, the ideal adaptive system immediately and exactly reproduces the phase of the reference wave in the entire cross section plane, including single points (dislocations) of the wavefront. In the second version, only the component corresponding to the potential part of the vector field of local wavefront tilts is corrected (see the model of an ideal phase difference sensor in Chapter 2). Hereafter this correction is called the "correction of potential (or vortex-free) phase." Thus, we have actually implemented four schemes of numerical experiment:

- an ideal compensation system,
- a system of compensation for the potential part of aberrations,
- an ideal PC system, and
- a PC system for the potential part of aberrations.

Four parameters of the problem are the length of the propagation path L, the lens diameter D, the wavelength λ, and the turbulence intensity C_n^2. According to Gurvich's similarity theory [16], the problem of propagation of a plane wave in the turbulent atmosphere is characterized by only two scales; for example, the coherence length r_0 as the transverse scale and the diffraction length at the coherence length $L_t = k r_0^2$ as the longitudinal scale. Then the problem is characterized by the normalized path length L/L_t and the normalized aperture diameter D/r_0. The scintillation index β_0^2 of the plane wave for the power spectrum of turbulence is unambiguously connected with the ratio L/L_t: $\beta_0^2 = 2.9(L/L_t)^{5/6}$. It can be used as a parameter instead of the ratio L/L_t.

The results of our numerical experiments are shown in Fig. 3.3.2. This figure shows the dependence of the Strehl ratio on the scintillation index for each correction scheme. The normalized aperture diameter took the values $D/r_0 = 10$, 20, and 30. Therefore, a family of three curves corresponds to each scheme.

For the ideal compensation scheme, the SR is nearly independent of either the normalized aperture diameter or the path length. The difference of the SR from the diffraction-limited value (equal to unity) is almost unnoticeable. This agrees with the idea that a major role in image distortion is played by phase fluctuations, whereas the role of amplitude fluctuations is insignificant [24-26]. In this case, the phase fluctuations are completely corrected.

In ideal compensation for only the vortex-free part of the phase, the correction efficiency decreases markedly as the scintillation index increases. The twofold decrease in the SR is achieved at a β_0^2 of about 1–1.5. As intensity scintillation increases further, the intensity at the lens focus tends to the uncorrected value. At $\beta_0^2 = 3$, i.e., $L = L_t = k r_0^2$, the correction efficiency decreases by an order of magnitude. Similar results have been obtained in the phase conjugation scheme (Fig. 3.3.3). This shows that application of this phase correction loses its efficiency when the path length achieves the diffraction length at the coherence length.

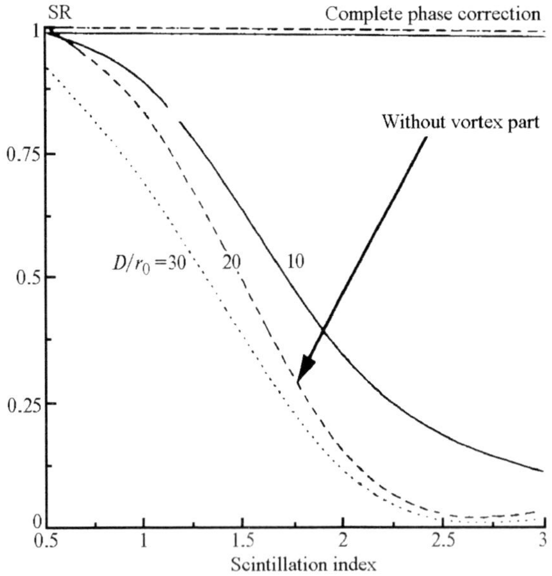

Figure 3.3.2. Dependence of SR on the scintillation index β_0^2 in the compensation scheme for phase distortions.

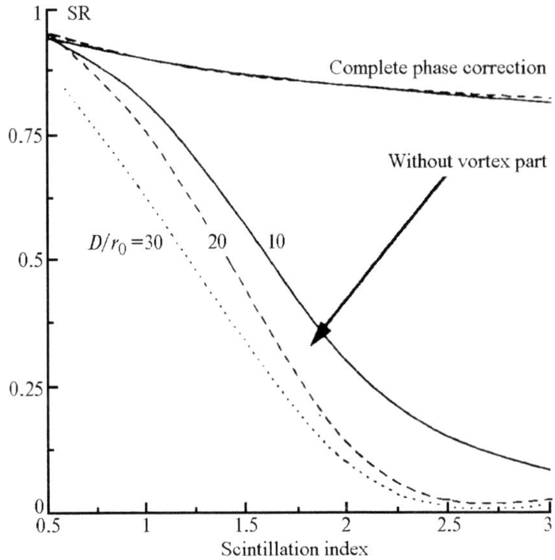

Figure 3.3.3. Dependence of SR on the scintillation index β_0^2 in the phase conjugation scheme.

The efficiency of correction in the ideal phase conjugation scheme is not as insensitive to the value of intensity scintillation. However, the dependence is not as strong as could be expected. At $\beta_0^2 = 3$, the SR decreases to 0.8 and does not depend on the aperture diameter.

There are experimental data with which we can compare the results of our numerical analysis. The experiment was conducted along a 5.5-km path [27] at Massachusetts Institute of Technology's Lincoln Laboratory. The adaptive system included a Hartmann sensor and a deformable mirror. The algorithm of phase conjugation for a focused beam was used. The wavelengths of the reference beam and the beam to be corrected were 633 and 514 nm, respectively.

Figure 3.3.4 shows the results of this experiment. The variance of fluctuations in the amplitude logarithm χ for a spherical wave is plotted along the horizontal axis. Note that the variance of intensity scintillation is four times larger than σ_χ^2 and the scales of the horizontal axes in Figs. 3.3.4 and 3.3.2 almost coincide.

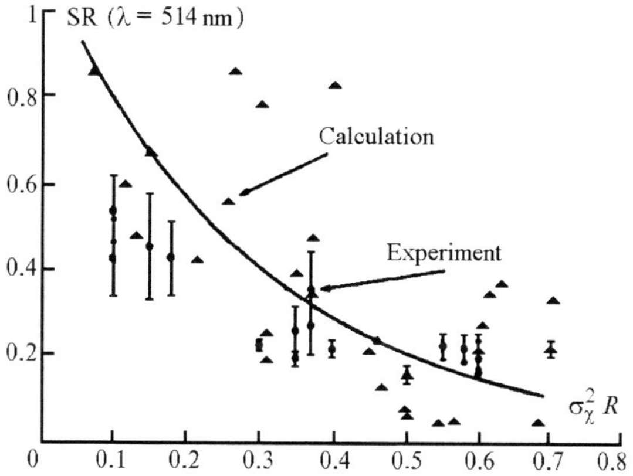

Figure 3.3.4. Experimentally obtained dependence of SR on the variance of amplitude fluctuations of a spherical wave and the dependence calculated [2, 16] in the Rytov approximation (method of smooth perturbations, MSP).

From the Lincoln Laboratory data, it is hard to judge to what degree the degradation of the correction efficiency is caused by phase discontinuity and filtering of phase dislocations, and to what degree it is caused by such factors as an increase in the ratio of the aperture diameter to the coherence length, errors of the wavefront sensor, and other errors of the adaptive system. The close agreement with our calculated data indicates that the use of the algorithm for reconstructing the wavefront of the reference beam that filters the "vortex" phase is the decisive factor causing degradation of the correction efficiency.

Based on this, the following conclusions can be drawn:

1. The proposed numerical model of an AOS that includes the "filtering" wavefront reconstruction algorithm allows adequate simulation and qualitative estimation of the efficiency of an existing adaptive system under conditions of strong intensity scintillation.
2. In the correction of turbulent distortions, the efficiency of phase correction roughly halves as the normalized variance of intensity scintillation (scintillation index) β_0^2 increases from zero to unity. In this range of β_0^2 values, the correction efficiency is nearly independent of the ratio between the aperture diameter and the coherence length. With a further increase in intensity scintillation, the dependence of the correction efficiency on the aperture diameter begins to manifest itself. An increase of β_0^2 up to 3 results in degradation of the correction efficiency by a factor of 10 and even more, and the SR tends to the value obtained in the system without correction.
3. Since the level $\beta_0^2 = 3$ roughly corresponds to the applicability boundary of the method of smooth perturbations (MSP), we can suppose that the applicability of the MSP is connected to the appearance of dislocations. Note that at the points of dislocations the intensity equals zero and the logarithm of the amplitude becomes infinite, whereas the MSP is in fact the perturbation method for the field logarithm.
4. The efficiency of adaptive correction of the vortex-free phase with the increasing variance of intensity scintillation degrades in roughly the same way in both the phase compensation scheme and the phase conjugation scheme. The differences between the plane wave and the beam are also insignificant, which follows from comparison with the experimental data obtained at the Lincoln Laboratory.

It is interesting to note that at strong amplitude fluctuations, the gain is nevertheless significant compared with the turbulent value of the SR. Table 3.3.1 gives the ratios of the corrected SR_c to the uncorrected SR_u for the compensation scheme for the vortex-free phase at $\beta_0^2 = 3$. It can be seen from Table 3.3.1 that at all D/r_0 the corrected value is roughly four times larger than the uncorrected one.

Table 3.3.1. Ratios of corrected and uncorrected SRc.

D/r_0	SR_u	SR_c	SR_c/SR_u
10	0.0324	0.129	3.98
20	0.0106	0.038	3.58
30	0.0051	0.025	4.90

References

1. M. Born, E. Wolf, *Optics Basics*, Nauka, Moscow, 1973.
2. V.I. Tatarskii, *Wave Propagation in a Turbulent Atmosphere*, Nauka, Moscow, 1967.
3. D.L. Fried, "Optical resolution through a randomly inhomogeneous medium for very long and very short exposures," *J. Opt. Soc. Am.*, **56**, No.10, pp. 1372–1379, 1966.
4. P.B.Vaganov, B.Z. Kacenelenbaum, *The Basics of Diffraction Theory*, Nauka, Moscow, 1982.
5. R.R. Beland, J.H. Brown, "A deterministic temperature model for stratospheric optical turbulence," *Physica Scripta*, **37**, pp. 419–423, 1988.
6. T. Stewart McKechnie, "Atmospheric turbulence and the resolution limits of large ground-based telescopes," *J. Opt. Soc. Am. A*, **9**, pp.1937–1954, 1992.
7. G.M. Bouricius, S.F. Clifford, "Experimental study of atmospherically induced phase fluctuations in an optical signal," *J. Opt. Soc. Am.*, **60**, pp. 1484–1489, 1970.
8. V.P. Lukin, V.V. Pokasov, "Optical wave phase fluctuations," *Appl. Opt.*, **20**, pp. 121–135, 1981.
9. D.L. Fried, "Statistics of a geometric representation of wavefront distortion," *J. Opt. Soc. Am.*, **55**, No. 11, pp. 1426–1435, 1965.
10. R.J. Noll, "Zernike polynomials and atmospheric turbulence," *J. Opt. Soc. Am.*, **66**, No. 3, pp. 207–211, 1976.
11. J.Y. Wang, J.K. Markey, "Modal compensation of atmospheric turbulence phase distortion," *J. Opt. Soc. Am.*, **68**, No. 1, pp. 78–88, 1978.
12. J. Winocur, "Modal compensation of atmospheric turbulence induced wavefront aberrations," *Appl. Opt.*, **21**, No. 3, pp. 433–438, 1982.
13. R.K. Tyson, "Adaptive optics system performance approximation for atmospheric turbulence correction," *Opt. Eng.*, **29**, No. 10, pp. 1165–1173, 1990.
14. V.P. Lukin, B.V. Fortes, "Partial correction of turbulent distortions in an ACT-10 telescope," *Atm. Oceanic Opt.*, **9**, No. 11, pp. 1492–1504, 1996.
15. V.P. Lukin, B.V. Fortes, "Efficiency of adaptive correction of images in a telescope using an artificial guide star," *OSA Tech. Digest*, **23**, pp. 92–93, 1995.
16. A.S. Gurvich, A.I. Kon, V.L. Mironov, S.S. Khmelevtsov, *Laser Radiation in a Turbulent Atmosphere*, Nauka, Moscow, 1976.
17. V.P. Lukin, B.V. Fortes, "Partial correction for turbulent distortions in a telescope," *Appl. Opt.*, **37**, pp. 4561–4568, 1998.
18. N.B. Baranova, B.J. Zel'dovich, "Dislocations of surface of wavefront and amplitude zeros," *JETF*, **80**, p. 1789, 1981.
19. D.L. Fried, J.L. Vaughn, "Branch cuts in the phase function," *Appl. Opt.*, **31**, No. 15, pp. 2865–2882, 1992.

20. V.P. Lukin, B.V. Fortes, "Influence of wavefront dislocations on phase conjugation instability at thermal blooming," *Atm. Oceac Opt.*, **8**, No. 3, pp. 435–447, 1995.
21. V.P. Lukin, B.V. Fortes, "Adaptive correction of turbulent distortions in conditions of strong intensity fluctuations. Influence of wavefront dislocations," in *Proceedings III Symposium on Atmospheric and Ocean Optics*, pp. 28–29, 1996.
22. B.V. Fortes, "Phase correction of turbulent distortions under strong intensity fluctuations," in *Proceedings IV Symposium on Atmospheric and Ocean Optics*, pp.186-187, 1997.
23. V.A. Banakh, V.L. Mironov, *Location of Propagation Laser Radiation in a Turbulent Atmosphere*, Nauka, Novosibirsk, 1986.
24. V.P. Lukin, B.V. Fortes, "Anisoplanatic degradation of correction with real beacon," *SPIE Proc. Vol.* **4338**, pp.107–117, 2000.
25. V.P. Lukin, B.V. Fortes, "Adaptive correction of focused beam in conditions of strong intensity fluctuations," *Atm. Ocean Optics*, 13, pp.515–520, 2000.
26. V.P. Lukin, B.V. Fortes, Adaptive phase correction of turbulent distortion for image and beam under conditions of strong intensity fluctuations," *Proc. SPIE Vol.* **4167**, pp. 180–191, 2000.
27. C.A. Primmerman, T.R. Price, R.A. Humphreys, B.G. Zollars, H.T. Barclay, J. Herrmann, "Atmospheric-compensation experiments in strong-scintillation condition," *Appl. Opt.*, **34**, No. 12, pp. 2081–2089, 1995.

CHAPTER 4

Minimization and Phase Correction of Thermal Blooming of High-Power Beams

Earlier we discussed examples of adaptive phase correction for laser beams and images formed in turbulent atmospheres. Thermal blooming, along with turbulent distortions, is a serious factor that limits concentration of the energy of laser beams on an object. In some situations, in particular, when using an IR laser, thermal blooming is almost the sole obstacle to achieving a high level of intensity on a target.

Various methods are used to make the transfer of energy of an optical beam more efficient. Relatively simple methods are the optimization of power, intensity distribution, focal length [1], and time conditions of radiation. The methods of a priori (programmed) phase correction (APC), phase conjugation, and wavefront reversal are more complex.

The first two sections in this chapter are devoted to the capabilities of optimization and a priori phase correction on vertical paths with allowance made for the altitude profile of the wind vector and the finite spatial resolution of the corrector in an adaptive system applied to beams with different initial intensity distributions in the cross section. The problem is considered in the approximation of a phase screen, and this allows us to estimate the correction efficiency for extremely wide beams on vertical paths.

Note that the effect of the initial intensity distribution in the beam cross section on the concentration of energy in the receiving aperture has been considered in a number of papers. For example, the problem of thermal blooming of a scanning beam was considered in Ref. [2] and it was shown that the homogeneous (like a super-Gaussian) intensity distribution gives better results than the Gaussian beam. Various beams having on-axis null and propagating along a homogeneous path without scanning were considered in Ref. [3]. It was shown that in this situation the non-Gaussian intensity distribution gives no significant gain. Nevertheless, for vertical paths such beams can give a marked gain [4].

The peculiarities of thermal blooming for hyper-Gaussian beams were considered in Ref. [5]. In other papers, square (hyper-Gaussian) and Gaussian beams were compared for a homogeneous (horizontal) path [6]. However, so far there has been no systematic study of thermal blooming for beams of different types propagating along a vertical path.

Note that an important peculiarity of vertical (and slant) atmospheric paths is the rather strong dependence of the wind direction on its altitude above the surface. In contrast to horizontal paths, this is caused by large-scale geophysical processes in the atmosphere, rather than random fluctuations of the wind velocity vector [7]. This factor significantly affects the aberration structure of phase distortions and consequently the efficiency of low-mode phase correction [8-10].

In the second half of this chapter we study the peculiarities of the phase conjugation algorithm for a homogeneous (horizontal) path, the causes for PC instability, and ways to increase the efficiency of this correction algorithm. The study of the efficiency of the PC algorithm for minimization of thermal blooming on a homogeneous path is a classical problem that was considered by many investigators dealing with adaptive correction of nonlinear distortions [11-15]. Here this problem is considered for the first time from the viewpoint of the effect of topological singularities in the phase of a reference beam (phase dislocations) on the correction process.

It turns out that the appearance of dislocations is closely connected with oscillations arising in a "fast" system of phase conjugation. A modification of modal phase conjugation is proposed that allows the correction efficiency to be noticeably improved.

In the following section we consider the problem of optimizing the power of the initial laser beam, first neglecting the altitude dependence of the wind direction and then estimating the effect of this dependence on widening of the focal spot of a high-power beam using a simplified model of the atmosphere. The problem is solved in the approximation of a phase screen; phase correction is ignored.

4.1 Amplitude Optimization for Thermal Blooming Along a Vertical Path

Consider the efficiency of energy transfer to an object when using coherent laser beams with the following intensity distributions:

1. Gaussian beam:

$$I(\vec{\rho}) = I_0 \exp\left[-\frac{\rho^2}{\left(\frac{1}{2}R\right)^2}\right] A(\vec{\rho}); \qquad (4.1.1)$$

$$I(\vec{\rho}) = I_0 \exp\left(-\frac{\rho^2}{R^2}\right) A(\vec{\rho}), \qquad \text{(wide Gaussian beam);}$$

2. Super-Gaussian beam:

$$I(\vec{\rho}) = I_0 \exp\left(-\frac{\rho^8}{R^8}\right) A(\vec{\rho}); \qquad (4.1.2)$$

3. Hyper-Gaussian beam:

$$I(\vec{\rho}) = I_0 \exp\left(-\frac{x^8 + y^8}{R^8}\right) A(x, y). \qquad (4.1.3)$$

The intensity I_0 can be determined from the normalization conditions:

$$\iint I(\vec{\rho})\, d^2\rho = P, \qquad (4.1.4)$$

where P is the total beam power. The aperture functions A were set so that the area of the emitting aperture was the same for all types of the beam and in addition, the intensity jump at the boundary of the emitting aperture was smoothed to some degree to avoid violation of the sampling theorem in numerical solution of the diffraction problem:

$$A(\rho) = \begin{cases} \exp\left(-\rho^{16}/\rho_0^{16}\right), & \rho \le R, \\ 0, & \rho > R \end{cases} \qquad (4.1.5)$$

for the Gaussian and super-Gaussian beams (R is the aperture radius);

$$A(x, y) = \begin{cases} \exp\left[-(x/\rho_0)^{16} - (y/\rho_0)^{16}\right], & |x| \le r \text{ and } |y| \le r \\ 0, & |x| > r \text{ or } |y| > r \end{cases}; \text{ here } (2r)^2 = \pi R^2$$

for the hyper-Gaussian beam. The value of ρ_0 was chosen so that at the inner part of the aperture boundary ($\rho = R$, or $x = r$, $y = r$) the aperture function took the value of 10^{-3}:

$$A(R) = \exp(-R^{16}/\rho_0^{16}) = 10^{-3}. \qquad (4.1.6)$$

4.1.1 Power optimization for beams with different cross-sectional intensity distributions

The intensity jump at the aperture boundary did not exceed 0.1%, and violation of the sampling theorem in numerical solution of the propagation problem was not important. Thus, the following versions of filling the emitting aperture were considered:

- a Gaussian beam whose efficient radius is equal to half the aperture radius,
- a wide Gaussian beam whose effective radius is equal to the aperture radius,
- a super-Gaussian beam almost uniformly filling the round aperture (round beam), and
- a hyper-Gaussian beam almost uniformly filling the square aperture (square beam).

For the hyper-Gaussian beam, two orientations of the square aperture were considered: a side (the first case) and a diagonal (the second case) of a square oriented along the wind direction.

Based on the known results calculated for non-Gaussian beams, we can assume that the gain is maximal if nonlinear distortions are concentrated maximally close to the emitting aperture. To increase the possible gain, consider the problem of thermal blooming on a phase screen located in the plane of the emitting aperture. The intensity distribution in the focal plane is determined in the approximation of the Fraunhofer diffraction, i.e., in the far diffraction zone. Thus the mathematical formulation of the problem is reduced to solution of the diffraction problem in the far zone with the boundary conditions in the following form:

$$U(\vec{\rho}, z=0) = \sqrt{I(\vec{\rho})} \times \exp[i\varphi(\vec{\rho})]. \tag{4.1.7}$$

The phase change $\varphi(\vec{\rho})$ on a nonlinear phase screen is determined from solution of the stationary equation of forced heat transfer:

$$\frac{\partial T}{\partial x} = \frac{\alpha}{\rho_0 C_P V_\perp} I(x, y), \quad \varphi(\vec{\rho}) = n_T' T \frac{2\pi}{\lambda} L, \tag{4.1.8}$$

which can be immediately written for the phase screen:

$$\frac{\partial \varphi}{\partial x} = I(x, y) \frac{k\alpha n_T' L}{\rho_0 C_p V_\perp}, \tag{4.1.9}$$

where L is the length of the nonlinear segment represented by the phase screen; α is the absorption coefficient of the medium, n_T' is the derivative of the refractive index with respect to the temperature T, $\rho_0 C_P$ is the product of the volume density and thermal capacity of the medium, and $V_\perp$ is the absolute value of the transverse component of the wind velocity vector. In dimensionless variables, this equation can be written as

$$\frac{\partial \varphi}{\partial x'} = I'(x', y'), \tag{4.1.10}$$

where the transverse coordinates are normalized to the effective radius of the Gaussian beam $a_0 = R/2$, and the intensity is normalized as

$$I'(x, y) = I(x, y) \frac{\alpha n_T' L}{\rho C_p V_\perp} k a_0 . \tag{4.1.11}$$

Normalization of the total beam power differs by the square linear scale; i.e.,

$$P' = P \frac{\alpha k n_T' L}{\rho C_p a_0 V_\perp} . \tag{4.1.12}$$

The results of optimization of the dimensionless power of different beams are shown in Fig. 4.1.1 as the power in a spot of diameter R_0:

$$R_0 = 0.64\ \lambda / D, \tag{4.1.13}$$

with R_0 being the effective angular size of the undistorted Gaussian beam at $D = 2a_0$.

It can be seen that non-Gaussian beams give a three- to fivefold gain as compared with Gaussian beams. The widened Gaussian beam truncated at the $1/e$ level at the boundary of the emitting aperture also gives a significant gain compared with the Gaussian beam truncated at the $1/e^2$ level. Apparently, the difference between beams with different intensity distributions is caused by two factors: first, different effective beam sizes, which lead to better results in the absence of nonlinear distortions, and second, different structures of aberrations arising in the nonlinear medium.

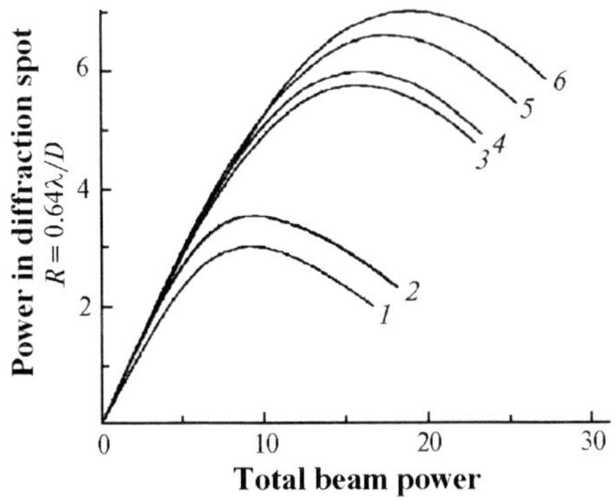

Figure 4.1.1. Dependence of the power in the diffraction spot on the total power of the beam (power optimization curves). Types of beams: (1) truncated Gaussian beam, (2) unlimited Gaussian beam, (3) hyper-Gaussian beam directed at an angle to the wind, (4) wide Gaussian beam, (5) super-Gaussian beam, and (6) hyper-Gaussian beam directed with a side to the wind.

4.1.2 Dependence of wind direction on the longitudinal coordinate

When a beam propagates along a path on which the wind direction depends on the longitudinal coordinate (this is typical of vertical and slant atmospheric paths), the structure of aberrations at the exit from the nonlinear medium also depends on the directional profile of the wind vector $\vec{V}_{\perp}(z)$. Therefore, the altitude dependence of the wind direction may be a significant factor affecting both the maximum achievable concentration of the energy on the object and the relation between beams of different types.

Combining various states of the atmosphere with various directions and scanning rates of the beam behind the object gives a very wide spectrum of dependencies of the vector of mutual displacement between the beam axis and the medium. Therefore, we will consider a simplified model of a path with the dependence of the wind direction on the longitudinal coordinate.

As before, we believe that nonlinear distortions are concentrated in a rather narrow layer near the emitting aperture, so their effect can be considered in the approximation of a phase screen. However, in this case, the phase screen is the sum of the phase screens corresponding to path layers moving in different directions. Denote now the phase distortions for the layer in which the angle between the vector of the transverse wind and the $0x$-axis is equal to β, as $\varphi(\vec{\rho}, \beta)$. These are described by the following equation in partial derivatives (primes of dimensionless variables are omitted):

$$\cos\beta\frac{\partial\varphi}{\partial x}+\sin\beta\frac{\partial\varphi}{\partial y}=I(\vec{\rho})w(\beta), \tag{4.1.14}$$

where $w(\beta)$ is the weighting function describing the angular distribution of nonlinear distortions and meeting the normalization condition:

$$\int_0^{2\pi} w(\beta)\; d\beta = 1\,. \tag{4.1.15}$$

The full-phase distortion at the exit from the nonlinear medium is now determined as

$$\varphi(\vec{\rho})=\int_0^{2\pi}\varphi(\vec{\rho},\beta)\; d\beta\,. \tag{4.1.16}$$

Formulation of the above problem in this section is a particular case corresponding to the limiting transition:

$$w(\beta)=\delta(\beta). \tag{4.1.17}$$

To describe this situation, it is convenient to use the first and second "moments" of the distribution $w(\beta)$:

$$\overline{\beta}=\int_0^{2\pi}\beta w(\beta)d\beta\;;\quad \sigma_\beta^2=\int_0^{2\pi}\left(\beta-\overline{\beta}\right)^2 w(\beta)d\beta\,. \tag{4.1.18}$$

Consider now the simplest case in which β is distributed homogeneously in some range $[-\Delta\beta/2,+\Delta\beta/2]$:

$$w(\beta)=\begin{cases}\dfrac{1}{\Delta\beta} & |\beta|\le\Delta\beta/2\\ 0 & |\beta|>\Delta\beta/2\end{cases}\;;\quad \Delta\beta=\sqrt{12}\sigma_\beta\,. \tag{4.1.19}$$

This model allows us to determine in a general way, both qualitatively and quantitatively, the possible limits of variability of the maximum achievable parameters of radiation on an object according to the profile of the effective transverse wind.

With the increasing width of the angular domain $\Delta\beta$, the structure of phase distortions becomes more symmetrical because of the combination of thermal lenses arising in the layers moving in different directions. Therefore, nonlinear refraction, astigmatism, and other asymmetrical aberrations are bound to decrease and, on the contrary, the contributions of total defocusing and other axisymmetrical aberrations to phase distortions are bound to increase.

Calculations showed that this leads to a considerable change in the optimal power. Figure 4.1.2 shows the power optimization curves for the Gaussian and hyper-Gaussian beams for $\Delta\sigma_\beta = 0.5$, 1, and 1.5 ($\Delta\beta \approx 100$, 200, and 300 deg). It is seen from Fig. 4.1.2 that symmetrization of an integral thermal lens arising from a combination of layers with different wind directions can result in an almost threefold increase of the power on the target. The following conclusion can be drawn from this consideration: The optimal power of a wide beam undergoing thermal blooming on a vertical path depends significantly on both the initial intensity distribution in the beam cross section and the altitude profile of the wind vector.

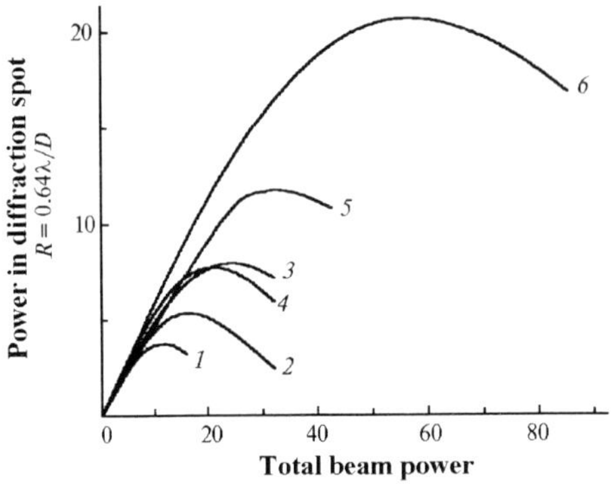

Figure 4.1.2. Power optimization curves for a path with the wind vector dependent on the longitudinal coordinate: (1, 2, 3) Gaussian beam, $\Delta\beta = 100$, 200, 300 deg; and (4, 5, 6) hyper-Gaussian beam, $\Delta\beta = 100$, 200, 300 deg.

4.2 Programmed Modal Phase Correction of Thermal Blooming Along a Vertical Path

Thermal blooming on a vertical path is a situation in which phase correction may be highly efficient [15] because nonlinear phase distortions are concentrated in a small (relative to the entire path) area near the emitting aperture, and the wave parameter of the effective length of this area is much less than unity for the wide beams that are usually used to transport radiation to extraterrestrial objects. In this situation, the main factor restricting the efficiency of phase correction of thermal blooming is the corrector's number of degrees of freedom. Let us now examine how the efficiency of programmed correction of the lowest modes of phase aberrations depends on the earlier described factors of cross-sectional intensity distribution of the beam along with the altitude dependence of the wind direction.

A priori (programmed) phase correction is based on information on the medium's parameters entering into the heat transfer equation. If they are known, then given the intensity distribution in the beam cross section, we can calculate the structure of the thermal lens in the beam channel. With the numerical representation of the 3D structure of optical inhomogeneities in the beam channel, we can select the initial phase profile of the optical wave at which the maximum, or close to the maximum possible, concentration of the energy on the object is achieved.

The efficiency of the a priori correction is restricted by the following factors:

- inaccuracy of the a priori information on the beam and path parameters affecting the formation of optical inhomogeneities in the beam channel,
- nonoptimality of the method used to construct phase correction, and
- finite spatial resolution of the correcting element of the adaptive system that limits the ability to obtain the exact formation of the phase profile.

The most significant components of the a priori information on the state of the atmosphere are first, the altitude profile of the wind velocity vector and then the less-variable profiles of the absorption coefficient and temperature. The principally inaccessible a priori information is the instantaneous state of turbulent inhomogeneities; this restricts the regions of the most efficient application of the APC to the far and midrange infrared.

The known methods of constructing phase correction can be presented as the following three groups in an order of increasing complexity:

1. Phase correction is determined as predistortion that is equal, with the opposite sign, to the integral phase distortion in the nonlinear medium:

$$\varphi_c(\vec{\rho}) = \frac{2\pi}{\lambda}\int \tilde{n}(\vec{\rho}, z)dz, \quad \tilde{n}(\vec{\rho}, z) = n(\vec{\rho}, z) - n_0 . \qquad (4.2.1)$$

   Inhomogeneities of the refractive index can be obtained here from solution of the thermodynamic part of the thermal blooming problem. This solution can be obtained from the beam intensity distribution in the plane of the emitting aperture [2, 16], from the 3D intensity distribution corresponding to the diffraction propagation of the beam [17], or by other methods.

2. Phase correction is determined from calculation of the wavefront of some reference radiation propagating along the channel of the high-power beam, but in the opposite direction [18]. This is essentially the method of phase conjugation, but with a mathematical (numerical) reference source in

place of the actual one. In contrast to a phase-conjugating adaptive system, the efficiency of this method depends on the accuracy of the a priori information used, as does the efficiency of any other APC method.
3. Phase correction is determined based on optimization of the coefficients of expansion on the basis of some functions [19], for example, on the basis of Zernike polynomials or the response functions of the phase corrector. In this case, the problem is to maximize the energy calculated by solving the thermal blooming problem with the available a priori information, rather than the actual concentration of the energy on the object.

Method (1) is simplest to implement. The phase correction is determined unambiguously and in a natural way, and we can expect that this method is less sensitive to errors in the a priori information. However, its efficiency may prove to be lower than the efficiency of methods (2) and (3), if the accuracy of the a priori information is high.

Consider the dependence of the APC efficiency on the corrector's configuration and the number of degrees of freedom. This allows one to determine the upper limit of correction efficiency with the use of currently available APC technologies. To exclude the effect of other factors on correction efficiency, we represent the nonlinear medium, as in the previous section, as a thin phase screen located in the plane of the emitting aperture. The phase correction is determined as a value equal to the nonlinear phase change:

$$\varphi_c(\vec{\rho}) = -\varphi_N(\vec{\rho}). \tag{4.2.2}$$

The corrector control can be found by minimizing the integral square error of approximation:

$$\sigma^2 = \iint d^2\rho [\varphi_c(\vec{\rho}) - \varphi(\vec{\rho})]^2 \to \text{min.} \tag{4.2.3}$$

Figure 4.2.1 shows the results calculated for a modal corrector and a Gaussian beam (at $\Delta\beta = 0$). Power normalization and other details of the numerical experiment are the same as in Sec. 4.1. One can see that a simple adaptive system that controls linear and quadratic aberrations allows the energy concentration on the object to be roughly doubled. Thus, a considerable increase in the energy concentration on the object is achieved at a relatively small number of degrees of freedom for the corrector. This is explained by the prevalence of the lowest (large-scale) aberrations in the structure of the thermal lens.

Now consider how the intensity distribution in the beam cross section and the dependence of the wind direction on the longitudinal coordinate affect the efficiency of the low-mode correction. Table 4.2.1 gives the power on the target (at the optimal beam power) for four different intensity distributions in the plane of the emitting aperture. The variable parameters were the angular width of the domain $\Delta\beta$ and the number of corrected aberrations. The first column gives the values obtained in the absence of correction; the second column gives those obtained with tilt correction; and the third column gives the values obtained with correction of tilt and defocusing. The values in the upper part of the table are normalized to the result for the uncorrected Gaussian beam at $\Delta\beta = 0$ deg, and those in the lower part have a different normalization for each $\Delta\beta$.

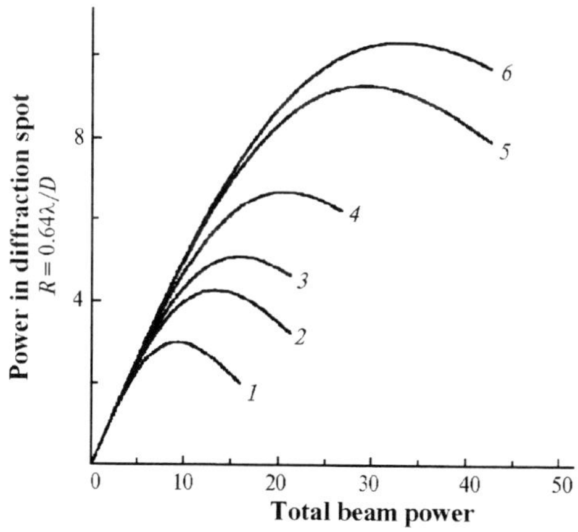

Figure 4.2.1. APC efficiency when using a modal corrector for a Gaussian beam: (1) no correction, (2) correction of tilt, (3) defocusing, (4) astigmatism, (5) coma, and (6) spherical aberration.

It can be seen that the dependence of the wind direction on the longitudinal coordinate along with a variable initial intensity distribution leads to a very wide variability in the efficiency of low-mode adaptive correction. For example, at a constant wind direction ($\Delta\beta = 0$), the correction of second-order aberrations (third column) gives a gain of only 1.67 times. At the same time, the correction of only tilts for the hyper-Gaussian beam allows the power to be increased almost 14 times. However, as $\Delta\beta$ increases, i.e., the dependence of the wind direction on the longitudinal coordinate becomes stronger, the advantage of the hyper-Gaussian beam decreases markedly, but the use of the super-Gaussian (round) beam in combination with correction of the first-order and second-order aberrations becomes very efficient (third column). The increased power on the target can achieve 37 times that for a Gaussian beam without correction and a constant wind direction.

Table 4.2.1. Power on the target at the optimal beam power.

Beam	Δβ								
	0 deg			150 deg			300 deg		
	I	II	III	I	II	III	I	II	III
Gaussian	**1**	1.44	1.67	1.44	2.33	3.56	2.67	2.78	6.67
Super-Gaussian	2.22	5.56	6.67	3.11	8.44	29.44	7.33	8.56	37.44
Hyper-Gaussian*	2.33	13.89	3.78	3	7.56	3.78	6.89	8	3.89
Hyper-Gaussian**	1.89	3.33	2.67	3	7.78	3.89	7.22	8.56	4
Gaussian	**1**	1.44	1.67	**1**	1.62	2.47	**1**	1.04	2.50
Super-Gaussian	2.22	5.56	6.67	2.16	5.86	20.4	2.75	3.21	14.0
Hyper-Gaussian*	2.33	13.89	3.78	2.08	5.25	2.62	2.59	3	1.46
Hyper-Gaussian**	1.89	3.33	2.67	2.08	5.4	2.70	2.7	3.21	1.5

* The side of the aperture is oriented along the direction of the mean wind.
** The diagonal of the aperture is oriented along the direction of the mean wind.

If the wind direction is independent of the longitudinal coordinate or if there is a dominant direction (for example, at scanning), the use of the hyper-Gaussian beam in combination with the tilt correction is most efficient; otherwise, the use of the super-Gaussian beam with the correction of defocusing is most efficient.

These results have been obtained under the assumption that nonlinear distortions are concentrated in the plane of the emitting aperture. In actuality, the thermal lens is extended and this distorts the additivity of correcting predistortions and nonlinear distortions of the wavefront; complete compensation cannot be achieved, even at the infinite spatial resolution of the corrector. Such a situation is characteristic of the horizontal path. In this situation, other phase correction algorithms are applied; for example, the algorithm of path conjugation that is considered in the next section.

4.3 Method of Phase Conjugation on a Horizontal Path

The PC algorithm is based on the reciprocity principle of propagation of electromagnetic waves. As applied to the problem of propagation of paraxial wave beams under consideration here, it can be illustrated most readily using the reciprocity of the parabolic wave equation for the slow component of the complex amplitude of electromagnetic oscillations in free space. If a reference beam with the boundary conditions

$$U(\vec{\rho}, z=L) = U_L(\vec{\rho}) \tag{4.3.1}$$

propagating from the plane $z = L$ to the plane $z = 0$ described by the equation

$$-2ik\frac{\partial U}{\partial z} = \Delta_{\perp} U \tag{4.3.2}$$

gives the following distribution of the complex amplitude:

$$U(\vec{\rho}, z=0) = U_0(\vec{\rho}), \tag{4.3.3}$$

then the beam with the boundary conditions

$$E(\vec{\rho}, z=0) = U_0^*(\vec{\rho}) \tag{4.3.4}$$

propagating from the plane $z=0$ in the positive direction of the $0z$-axis to the plane $z = L$ (propagation is diffraction limited) and described by the equation

$$2ik\frac{\partial E}{\partial z} = \Delta_{\perp} E, \tag{4.3.5}$$

gives the distribution of the complex amplitude conjugate with the initial field of the reference beam:

$$E(\vec{\rho}, z=L) = U_L^*(\vec{\rho}). \tag{4.3.6}$$

Equation (4.3.4) describes a wavefront reversal correction algorithm. In this case, the wave E is the reversal of the wave U, and at every space point the Umov-Pointing vector of the wave E is equal in the absolute value and opposing in the direction to the Umov-Pointing vector of the wave U. The reciprocity principle is valid not only for free space but also for an optically inhomogeneous medium; therefore the WFR algorithm can be used to correct atmospheric distortions.

Although some optical devices based on nonlinear interactions of optical waves in crystals are capable of implementing (with a limited accuracy) [20] the WFR described by Eq. (4.3.4), in practice they usually restrict their operation to phase control (phase conjugation) corresponding to the equation

$$E(\vec{\rho}, z) = \sqrt{I(\vec{\rho})}\exp[-\arg U(\vec{\rho}, 0)], \tag{4.3.7}$$

where $I(\vec{\rho})$ is the initial intensity distribution of the wave to be corrected.

Thus, the information borne by the amplitude of the reference wave U is lost in phase conjugation. In Chapter 3 it was shown that the loss of this information does not lead to a significant decrease in the efficiency of correction of turbulent distortions. However, under conditions of nonlinear interaction of the radiation with the medium, phase correction leads to specific effects, including instabilities of a different character. The main problem considered in this section is the mechanism of development of PC instability during thermal blooming of a focused beam on a homogeneous horizontal path.

The papers by Herrmann [11, 21] were most likely the first publications reporting the possibility of an unstable PC mode. In Ref. [11] with the simplified mathematical model of nonlinear propagation, it was shown that starting from some threshold power, the application of phase conjugation results in an unbounded growth in divergence of the beam to be corrected. It was also mentioned that small-scale instability was observed in the numerical experiment.

The mechanism of development of small-scale instability has been considered in detail in recent papers [22, 23]. This type of instability is most likely characteristic of the situation when nonlinear (and possibly turbulent) distortions are concentrated within a short section of a path whose length is much shorter than the diffraction length of the beam. This is typical of vertical atmospheric paths.

Horizontal paths are characterized by instability of the oscillatory type [15, 16]. However, the mechanism of its development was not fully understood. As we have shown in Ref. [24], this instability is connected with wavefront dislocations in the reference beam.

The problem of adaptive correction is considered in the following formulation: A high-power focused beam propagates in a homogeneous medium in which there are no random inhomogeneities and all parameters determining thermal blooming are constant along the entire path. The reference beam propagates from the focal plane in a direction opposite to the high-power beam along its channel (its axis) without thermal blooming. The boundary conditions for the reference beam correspond to the WFR of the diffraction-limited high-power beam. In the plane $z = 0$ (on the emitting aperture of the adaptive system), the phase distribution of the corrected high-power beam at every moment of time is equal to the phase distribution in the reference beam with the opposite sign. This corresponds to the algorithm of fast phase conjugation. Mathematically, the problem is formulated by the following set of equations:

$$2ik\frac{\partial E}{\partial z} = \Delta_{\perp} E + 2k^2 \tilde{n} E, \quad E(\vec{\rho}, z=0, t) = \sqrt{I(\vec{\rho})}\exp[i\varphi(\vec{\rho})],$$

$$-2ik\frac{\partial U}{\partial z} = \Delta_{\perp} U + 2k^2 \tilde{n} U; \quad U(\vec{\rho}, z=f) = E(\vec{\rho}, z=f, t=0), \qquad (4.3.8)$$

$$\varphi(\vec{\rho})=\begin{cases} k\rho^2/(2f) & t=0 \\ -\arg U(\vec{\rho},z,t) & t>0 \end{cases}; \quad I(\vec{\rho})=I_0\exp\left(-\frac{\rho^2}{a_0^2}\right),$$

$$\frac{\partial\tilde{n}}{\partial t}+V_\perp\frac{\partial\tilde{n}}{\partial x}=\frac{\alpha n_T'}{\rho_0 C_P}EE^*.$$

In the focal plane, the peak intensity I_{max} and the power P in a circle of radius $R=L/(ka_0)$ corresponding to the effective radius of the diffraction-limited focused Gaussian beam were measured. The transition to the dimensionless variables gives the following equations for the normalized intensity and power:

$$I'=I\frac{\alpha n_T' k^2 a_0^3}{\rho_0 C_p V_\perp}; \quad P'=P\frac{\alpha n_T' k^2 a_0}{\rho_0 C_p V_\perp}. \tag{4.3.9}$$

The calculated results shown in Figs. 4.3.1–4.3.3 are given in this normalization. The time t is normalized to the convective time $2a_0/V$. The focal length was set equal to $f=\sqrt{10}\,ka_0^2$.

As the power increases above some threshold, the well-known oscillations [15] arise in the parameters of the corrected beam. To understand the mechanism of the appearance of these oscillations, let us examine the behavior of the waist of the corrected beam. To do this, we record the waist position as a distance between the plane of the emitting aperture and the plane in which the peak intensity in the beam cross section reaches the maximum. Because the problem is solved numerically, this coordinate z_{max} is determined accurately to the discretization step Δz along the longitudinal coordinate.

Figure 4.3.1 shows the dynamics of the waist position (a) and the peak intensity (b) in the waist cross section for the power at which oscillations are observed in the system. As follows from a comparison of this figure with Fig. 4.3.2, the oscillation period of the waist position corresponds to the oscillation period of the focal spot parameters.

Initially, the waist is situated near the focal plane, which follows from the theory of diffraction of a focused Gaussian beam. Then the waist shifts toward the emitting aperture as a result of correction of defocusing introduced by the evolving thermal lens. The reference beam passing through the thermal lens formed as a result of thermal blooming of the high-power beam to be corrected experiences defocusing. As a result of application of the phase conjugation algorithm, this defocusing transforms into extra focusing of the beam to be corrected. This leads to a waist displacement toward the APC aperture and a corresponding decrease in the system's focal length.

The diffraction-limited width of the beam decreases proportionally to the focal length, whereas the intensity has a square dependence. Therefore, the decrease in the focal length leads to a significant increase in the intensity at the waist. The intensity increase causes the corresponding growth of the temperature gradient in the thermal lens, leading to a further increase in defocusing of the reference beam and focusing of the beam to be corrected.

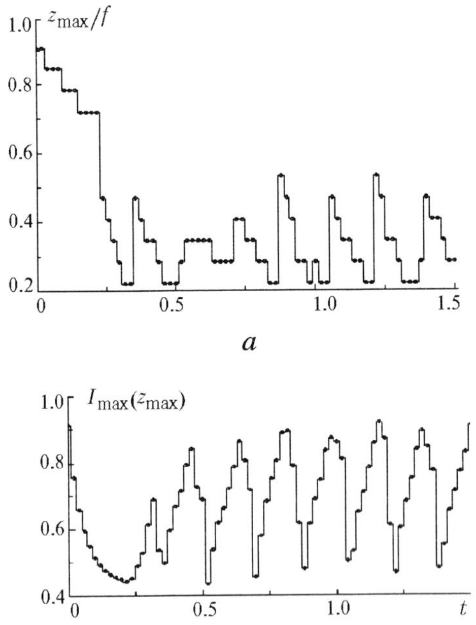

b

Figure 4.3.1. Dynamics of (a) the waist position and (b) the peak intensity in the waist cross section at I_0=24. The intensity is normalized to the diffraction-limited value of the intensity at the beam focus. The time t is normalized to the convective time $2a_0/V$.

Thus, a positive feedback is formed in the optical loop of the adaptive system. This feedback can be represented as the following chain: appearance of a thermal lens, defocusing of the reference wave, focusing of the high-power beam, shift of the waist to the emitting aperture, increase in the intensity at the waist, and growth of the thermal lens. However, no unbounded increase in the "signal" is observed in this chain. Consequently, there are some factors limiting the waist shift and leading to saturation of this nonlinear system with the feedback. The following factors most likely act simultaneously in this case:

1. The shift of the waist depends on the distance between the AOS emitting aperture and the part of the path where the thermal lens is concentrated, i.e., on the waist position at the previous iteration. The closer to the AOS aperture this part is, the smaller the waist shift, because as the beam passes through the waist, the excess focusing is compensated for by the thermal lens.

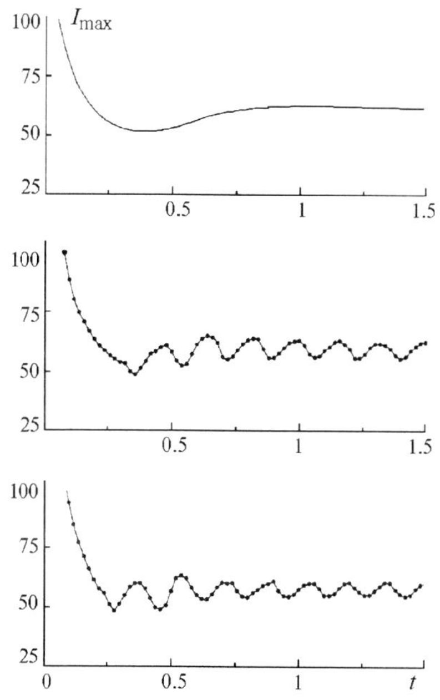

Figure 4.3.2. Dynamics of peak intensity at the focus of the beam to be corrected for different values of the initial intensity: $I_0 = 16$ (upper panel), 24 (central panel), and 32 (lower panel).

2. As the waist shifts and the effective size of the beam in this cross section decreases, the effective cross size of the thermal lens decreases too. At the same time, the size of the reference beam does not decrease to the same degree. Therefore, the thermal lens affects only the central part of the reference beam, and consequently only the central part of the beam to be corrected undergoes extra focusing.
3. Common defocusing prevails at the initial stage of formation of the thermal lens. Since the time $t > (2-3)a/V_{\perp}$ (where a is the effective size of the beam in the cross section), defocusing in the direction perpendicular to the wind direction, i.e., astigmatism, begins to prevail. The corresponding focusing of the beam to be corrected at this axis leads to a decrease in the beam size in only one direction, and the intensity at the waist decreases proportionally to the corresponding focusing parameter rather than its square.

Thus, there are at least three factors limiting an increase in focusing in the AOS considered. This indeed does not explain why the fast AOS turns into a self-oscillation regime, nor does it explain why this regime starts from some threshold power.

Let us now try to find the cause for oscillatory instability in the behavior of the reference wavefront. It is known that the WF topology can acquire a qualitatively new character at sufficiently strong distortions [25]. WF dislocations at the intersection of lines of sign alternation in the real and imaginary parts of the complex amplitude impart the character of a multilobed surface to the wavefront, and this surface projected onto the 2D cross section gives a phase distribution with discontinuity lines at 2π.

It is logical to suppose that a qualitatively new character of the topology of the reference wavefront can lead to a qualitatively new (self-oscillation) regime of AOS operation. Consider the results of a study of the WF topology of the reference beam [24]. The study was performed in the following way: At every iteration of the fast phase conjugation, the field of the reference beam was checked for dislocations and the coordinates of dislocations were fixed. Detection of dislocations was based on the well-known fact that the loop integral of the phase gradient

$$\oint \vec{g}d\vec{l} = (N_+ + N_-)2\pi \tag{4.3.10}$$

is proportional to the sum of the number of positive N_+ (i.e., twisted in the positive direction) and negative N_- topological charges inside the loop. A square whose vertices lie in four neighboring nodes of the computational grid is the smallest loop in numerical simulation. Since the integral along the straight line between the vertices is equal to the phase difference between these two points, the loop integral can be calculated as

$$\Sigma = \Delta_1 + \Delta_2 + \Delta_3 + \Delta_4 , \tag{4.3.11}$$

where $\Delta_1, ..., \Delta_4$ are the phase differences between the square vertices calculated in the following way: $\Delta_1 = \arg\,(U_2 U_1^*)$, and so on, where U_1 is the complex amplitude at the vertex 1 and so on. A dislocation was considered detected if $|\Sigma| > \pi$. In most cases, $|\Sigma| \approx 0$ was obtained accurately to a rounding error for mathematical operations. Coordinates of the loop center were assigned to the dislocation found in this way.

It turned out that two dislocations with opposite signs arise at the power corresponding to the threshold of the self-oscillation regime in the reference beam. Dislocations move in the direction of the wind vector and go beyond the recording zone. The time dependence of the dislocation coordinate is shown in Fig. 4.3.3.

Comparison of this plot with the plots for the parameters of the waist and the focal spot shows that the periodicity of all three processes is the same. This means that these oscillations are closely connected with the appearance of dislocations in the reference beam. The comparison of all three figures forms the basis for the following interpretation of the AOS oscillatory regime:

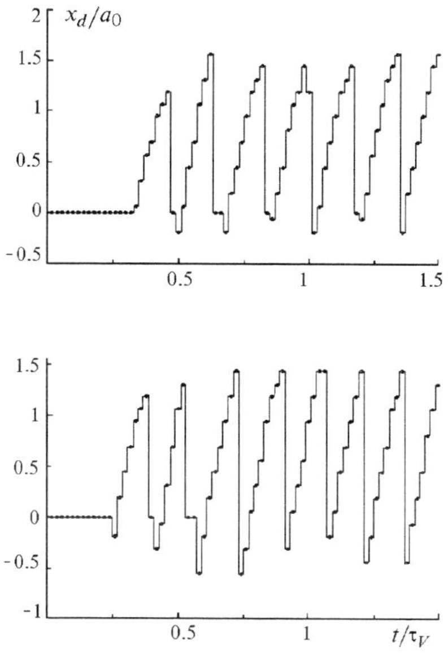

Figure 4.3.3. Time dependence of the position (X_d) of dislocation: I_0=24 (upper panel) and 32 (lower panel).

The shift of the waist of the high-power beam and the corresponding increase in its intensity lead to a strong local thermal lens. When distortions of the reference beam become strong enough and dislocations arise in it, the information about the thermal lens that is contained in the phase of the reference radiation is distorted so strongly that the optical feedback loop breaks.

Then the processes in the AOS evolve most likely in the following way: The local thermal lens situated in the cross section where the beam waist was located moves in the wind direction according to the mechanism of forced heat transfer. This process is accompanied by dislocations of the reference beam in the same direction. The motion is observed in the plane of the AOS aperture. Once the local thermal lens goes beyond the effective cross section of the reference beam and the corresponding wavefront dislocations go beyond the AOS aperture, the phase optical feedback is restored, and this leads to a new cycle of focusing growth.

Phase conjugation is WFR minus the information about amplitude distortions. Therefore, to the greater degree that phase distortions are transformed into amplitude distortions, the lesser the efficiency of the phase correction. When distortions of the amplitude profile become so strong that the intensity reduces to zero at some points, a qualitatively new situation arises. If the difference between the WFR and PC is characterized by the intensity ratio of the beam to be corrected and the reference beam at every point of the AOS aperture (for the WFR it is equal to unity with the corresponding normalization), then the appearance of dislocations makes this parameter infinite at the points where dislocations arise.

Apparently, in the problem considered we are dealing with exactly this mechanism of breaking of the phase optical feedback located in the functioning of the AOS. This break is so significant that focusing not only ceases to grow, but decreases rapidly, as evidenced by the quick shift of the waist to the focal plane.

4.4 Modal Phase Conjugation on a Horizontal Path

The model of an adaptive optical system considered in Sec. 4.3 corresponds to some hypothetical adaptive system that transfers the phase of the reference wave into the correcting beam with infinite spatial and temporal resolution or at least a resolution no less than the resolution of the numerical model used. First, we consider the efficiency of modal phase conjugation for propagation of a horizontal path optical beam.

In actuality, the spatial scales of existing AOSs are finite. The results obtained correspond to the so-called zonal principle of phase correction, according to which the relative phase is measured at some set of points of the AOS aperture and the wavefront of the emitted wave is corrected in a similar way. In our case, this set of points corresponds to the nodes of the computational grid.

4.4.1 Calculation of the efficiency of modal phase conjugation

Another approach is known; it is a modal approach and its phase surface is determined by the vector of the coefficients of phase expansion on some basis. Since only the large-scale part of the phase expansion (corresponding to the lowest aberrations) is used in practice, small-scale local wavefront deformations are not reconstructed in this AOS. It is logical to suppose that filtering of the small-scale part of phase correction can considerably affect the AOS operation at a beam power exceeding the threshold of the self-oscillation regime. As shown earlier, such oscillations are connected with the appearance of wavefront dislocations having a local (small-scale) character.

Since the 2D phase distribution in the presence of wavefront dislocations is determined with an uncertainty greater than a constant or even total wavefront tilt, the formulation of the problem of a modal PC (hereafter, an MPC) is intimately connected not only with the expansion basis used, but also with the specific type of the sensor and the algorithm for estimating WF. The modal algorithms for estimating WF are most often applied to sensors measuring a local phase gradient or wavefront tilt. A typical representative of such devices is the Hartmann sensor.

In the following discussion, we present the results of numerical simulation of phase correction of thermal blooming in a phase conjugation AOS with the Shack–Hartmann sensor. The parameters of the high-power and reference beams as well as the propagation path are the same as in Sec. 4.3. The technique of simulation of the wavefront sensor and the modal algorithm used to estimate aberrations of the reference beam phase are described in Chapter 2.

We simulated an AOS with a Shack–Hartmann sensor consisting of 16 subapertures (Fig. 4.4.1). Four corner subapertures were neglected, and local tilts were estimated in only 12 subapertures. The diameter of the sensor aperture $D = 4a_0$ corresponded to the diameter of the beam at the $1/e^2$ intensity level. We considered the scheme of a closed loop in which the reference wave passed through the same optical elements as the corrected beam. These elements were the focusing system and the modal corrector. The sensor measured only residual aberrations, i.e., the phase of the reference beam minus the diffraction component and distortions compensated for by the corrector at the previous iteration of phase conjugation. Thus, the field of the reference beam on the entrance pupil of the sensor U_S can be written as

$$U_S(\vec{\rho},t) = U(\vec{\rho},0,t) \times \exp\left[i\frac{k\rho^2}{2f} + i\Phi(\vec{\rho},t-\Delta t) \right], \qquad (4.4.1)$$

where $U(\vec{\rho},0,t)$ is its complex amplitude in the plane of the AOS receiving–emitting aperture at the current time, f is the focal length of the forming telescope, Δt is the time between PC iterations (correction delay time), and Φ is phase correction.

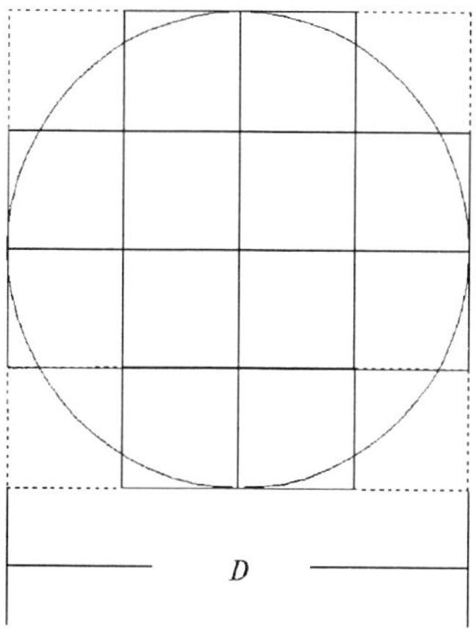

Figure 4.4.1. Configuration of the Shack–Hartmann sensor.

Phase correction was determined as a sum of the first 15 Zernike polynomials Z_l :

$$\Phi(\vec{\rho},t)=\begin{cases}\Phi(\vec{\rho},t-\Delta t)+\Delta\Phi, t\geq\Delta t \\ 0, \qquad t<\Delta t\end{cases}; \quad \Delta\Phi=\sum_{l=2}^{15}c_l(t)\times Z_l\left(\frac{\vec{\rho}}{2a_0}\right), \tag{4.4.2}$$

with the weights c_l obtained from modal estimation of the phase on a circle inscribed in the sensor aperture.

Numerical experiments with such AOS models, which have been conducted by us for the first time [24], have shown that correction of thermal blooming using an MPC is characterized by higher stability. This is illustrated in Fig. 4.4.2, which shows the dynamics of the peak intensity in the focal plane for different initial beam intensities (I_0 = 24, 32, and 64). In all the cases, oscillations are absent, i.e., the application of such a sensor either dampens them or at least increases their threshold.

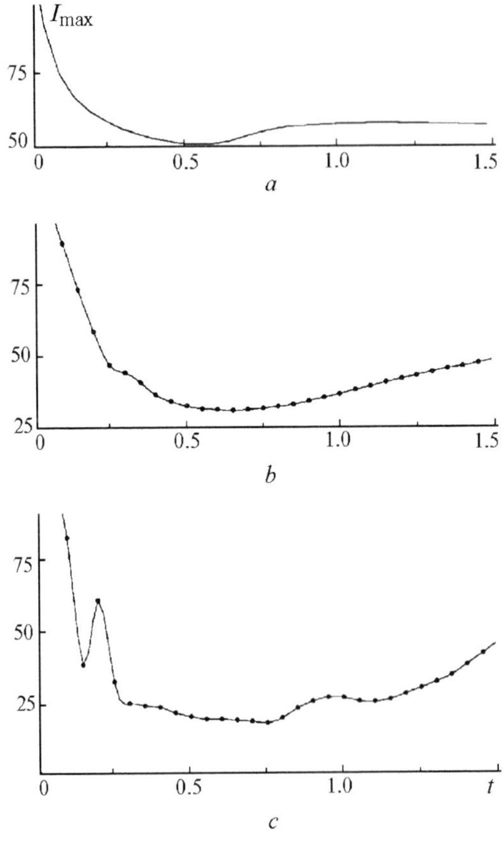

Figure 4.4.2. Dynamics of the peak intensity in the focal plane at I_0 = 24 (a), 32 (b), and 64 (c) for MPC.

A study of the reference beam has shown that dislocations arise in it, but this does not lead to development of AOS instability. The cause of the transition to the steady state can be interpreted as filtering of local deformations of the reference wavefront when the modal algorithm for estimating aberrations of the reference wave is applied.

In the AOS steady state, dislocations of the reference wavefront, if any, either disappear or keep their position (Figs. 4.4.3 and 4.4.4). It is interesting to note that at the focus of the sensor subaperture within which a dislocation falls, the focal spot splits into two halves. The evolution of such a double focal spot is illustrated in Fig. 4.4.5. An analogy of this effect for an interferometric device is known [25]. In the latter case, the appearance of a dislocation leads to splitting of an interferometric fringe, which acquires a forked shape. We can observe a similar effect in the AOS considered if the bias interferogram of the reference wave is recorded in parallel with phase correction.

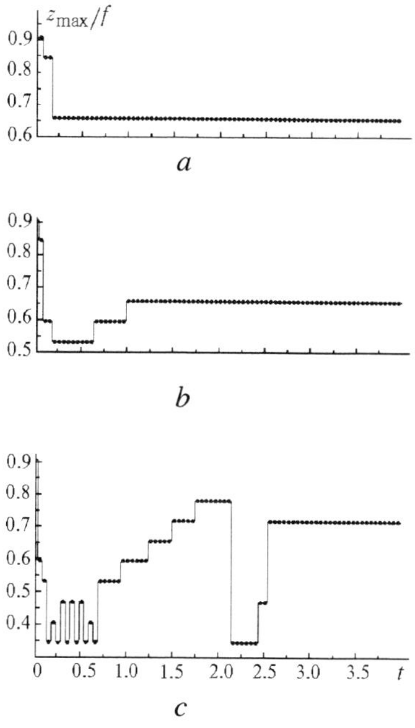

Figure 4.4.3. Waist position (MPC) at $I_0 = 24$ (a), 32 (b), and 64 (c).

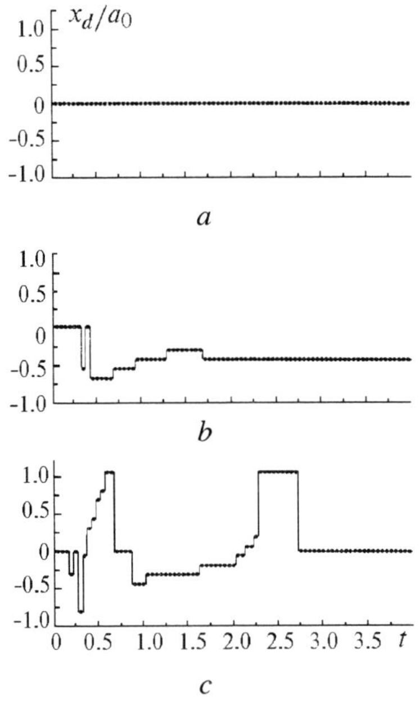

Figure 4.4.4. Position of WF dislocation in the reference beam at $l_0 = 24$ (a), 32 (b), and 64 (c).

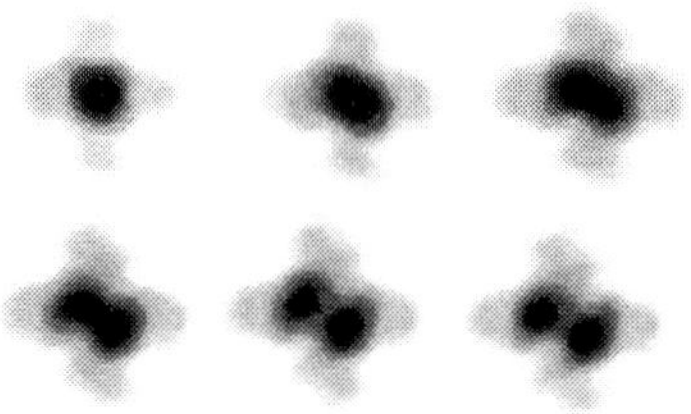

Figure 4.4.5. Appearance of a double focal spot at the focus of the subaperture of the Hartmann sensor.

4.4.2 Modification of modal phase conjugation

The waist shift to the aperture plane observed in the numerical experiment is, generally speaking, an unwanted effect because it leads to formation of a stronger thermal lens than in thermal blooming without correction. The waist shift is caused by the AOS tendency to compensate for thermal defocusing. This leads to a decrease in the focal length of the system or, what is the same, to an increase in the focusing parameter. It is known that the optimal value of the focusing parameter on a homogeneous path with thermal blooming is less than that for a diffraction-limited beam [4]. The nature of this effect is the same: an increase in the focusing parameter causes a waist shift toward the emitting aperture and the formation of higher temperature gradients in the regions adjacent to the waist.

Since the phase correction in an MPC is set as a weighted sum of Zernike polynomials, it is possible to directly control the phase correction modes. To prevent an unwanted waist shift, we can turn off the focusing control. In the mathematical description of the AOS loop, this corresponds to the following modification of the equation of phase correction:

$$\Delta\Phi(\vec{\rho},t)=\sum_{l=2}^{15}(1-\delta_{l,4})c_l(t)\times Z_l\left(\frac{\vec{\rho}}{2a_0}\right),\quad \delta_{lm}=\begin{cases}0, l\neq m\\ 1, l=m\end{cases}. \tag{4.4.3}$$

Since defocusing along the axis perpendicular to the wind vector, i.e., astigmatism, becomes the dominant quadratic aberration on completion of the transient process, we can try to turn off the astigmatism control as well:

$$\Delta\Phi=\sum_{l=2}^{15}(1-\delta_{l,4})(1-\delta_{l,5})(1-\delta_{l,6})\times c_l Z_l\,. \tag{4.4.4}$$

Numerical experiments have shown that removal of quadratic aberrations actually leads to an increase in the concentration of the beam energy on the target, and turning off both the focusing and astigmatism control gives the best result. This correction will be referred to as modified modal phase conjugation (MMPC). In addition to quadratic aberrations, tilt and coma contribute considerably to distortion of the beam phase under thermal blooming; therefore, MMPC mostly leads to directing the beam and straightening the characteristic "sickle" caused by coma.

Compare the efficiency of MMPC, MPC, and PC using the power optimization curves as a basis. Figure 4.4.6 shows the steady-state power in the diffraction spot P_d as a function of the initial intensity of the beam to be corrected for MPC (with correction of all lowest aberrations up to and including the fourth order) and MMPC, i.e., without control of quadratic aberrations, as well as with a complete PC and in a system without correction. The data for PC have been obtained by time averaging of the corresponding instantaneous values for several periods of oscillation.

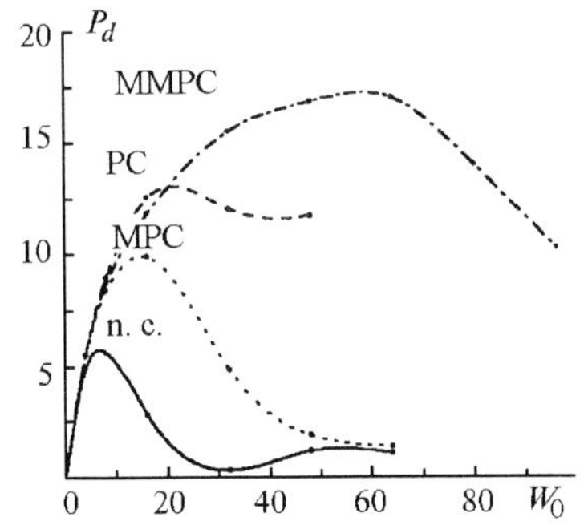

Figure 4.4.6. Power in the diffraction spot P_d as a function of the initial intensity I_0 at no correction (n.c.) and different correction algorithms (MMPC, MPC, and PC).

It can be seen that the MPC efficiency is somewhat lower than that of the PC, whereas the MMPC is far more efficient than a PC starting from the power at which dislocations and a self-oscillation regime arise ($I_0 \approx 20-24$). In combination with optimization of the initial beam intensity, the MMPC gives an almost threefold gain in the power P_d compared with the system without correction, and gains of 1.5 and 1.7 times over the PC and MPC, respectively.

We do not claim that the proposed modification of phase conjugation has versatility. It only gives a key to optimizing the correction algorithm by introducing changes in the control channels for such aberrations, which lead to unwanted increases in temperature gradients in the channel of the high-power beam. Perhaps the value of the focusing parameters in many cases should be set somewhat less than that optimal for diffraction-limited propagation. Under conditions of phase correction of thermal blooming in a turbulent medium, it makes sense to limit the range of correction of quadratic aberrations, rather than turn off the control of these aberrations.

Summarizing the discussion of modal phase conjugation in correcting for thermal blooming, we would like to note the main results obtained in this section. First, it has been shown that the appearance of dislocations in the MPC does not lead to the self-oscillation regime. At any power at which we succeeded in conducting the numerical experiment, the system always reached the steady state. We can suppose that the threshold for appearance of oscillatory instability becomes far higher. In the problem considered, it was higher than the optimal beam power. The absence of the self-oscillation regime indicates that no rigid requirements need be imposed on the AOS speed if the level of turbulent distortions is low.

Second, we have proposed a modified MPC that allows the efficiency of phase correction to be improved. This improvement is caused by the smaller degree of distortion of the amplitude profile of the reference beam that is due to preventing the development of strong temperature gradients in the channel of the high-power beam as the waist shifts toward the AOS aperture.

Thus, in this chapter we have considered two limiting geometries: a nonlinear phase screen at the beginning of the propagation path and a homogeneous path whose nonlinear properties are the same over the entire length from the emitting aperture to the beam focus. These two limiting cases have pronounced peculiarities in the problem of thermal blooming of a high-power focused beam.

In reality, various intermediate situations occur in the atmosphere, and peculiarities of one or another propagation scenario manifest themselves to a greater or lesser degree, depending on the direction of the beam axis relative to the horizon, the angular scanning rate of the optical system, the path altitude, and other factors. In general, it is difficult to recommend particular methods and devices of adaptive correction, but the regularities obtained here can serve as guidelines for specific problems.

References

1. V.V. Kolosov, S.I. Susuev, "Analysis of some algorithms of minimization of angular spreading for partially coherent optical radiation," *Atm. Opt.*, **3**, No. 1, pp. 83–89, 1990.
2. L.C. Bradley, J. Herrmann, "Phase compensation for thermal blooming," *Appl. Opt.*, **13**, No. 2, pp. 331–334, 1974.
3. J.E. Pearson, C. Yeh, W.P. Brown, "Propagation of laser beams having an on-axis null in the presence of thermal blooming," *J. Opt. Soc. Am.*, **66**, No. 12, pp. 1384–1388, 1976.
4. P.A. Konyaev, V.P. Lukin, "Thermal blooming of focused laser beams in the atmosphere," *Izv. V.U.Z. Fiz.*, **XXVI,** No. 2, pp. 79–89, 1983.
5. Yu.N. Karamzin, A.P. Sukhorukov, V.A. Trophimov, "Nonlinear distortions of hyper-Gaussian beams," *Izv. V.U.Z. Fiz.*, **XXVII**, No. 10, pp. 1292–1298, 1984.
6. J.D. Weiss, W.H. Maclnnis, "Thermal blooming: Round beam vs. square beam," *Appl. Opt.*, **19**, No. 1, pp. 31–33, 1980.
7. L.G. Matveev, *Course of General Meteorology*, Gydrometeoizdat, Leningrad, 1976.
8. P.A. Konyaev, V.P. Lukin, B.V. Fortes, "Phase correction of nonlinear distortions for a coherent beam," *Atm. Opt.*, **1,** No. 4, pp. 71–75, 1988.
9. P.A. Konyaev, V.P. Lukin, B.V. Fortes, "About phase correction of nonlinear distortions for focusing of coherent optical radiation through the atmosphere," *Atm. Opt.*, **3**, No. 12, pp. 1279–1285, 1990.
10. P.A. Konyaev, V.P. Lukin, B.V. Fortes, "Efficiency of program phase correction on vertical atmospheric paths," *Proceedings XV All-Union Conference on Wave Propagation,* Alma-Ata State Univ., p. 379, 1987.

11. J. Herrmann, "Properties of phase conjugate adaptive optics systems," *J. Opt. Soc. Am.*, **67**, No. 3, pp. 290–295, 1977.
12. M.A. Vorontsov, "Method of phase conjugation in the problem of thermal blooming compensation for light fields," *Sov. J. of Quantum Electron.*, **6**, No. 10, pp. 2078–2083, 1979.
13. V.V. Vorob'ev, "Estimations of possibilities of laser beams thermal blooming compensation with phase conjugation adaptive systems," *Ser. Radiotecnika i Electronika, Izv. AS USSR*, **26,** No. 11, pp. 2334–2341, 1981.
14. M.A. Vorontsov, V.P. Sivikon', V.I. Schmal'gauzen, "Method of phase conjugation for optical beaming adaptive system," *Izv. V.U.Z. Ser. Fiz.*, No. 3, pp. 26–37, 1983.
15. V.L. Mironov, P.A. Konyaev, V.P. Lukin, "The efficiency of application of adaptive optical systems under radiation self-action in the atmosphere," *Izv. V.U.Z. Ser. Fiz.*, **28**, No. 3, pp. 536–540, 1985.
16. V.E. Zuev, P.A. Konyaev, V.P. Lukin, "Minimization of atmospheric distortions of optical waves with adaptive optics methods," *Izv. V.U.Z. Ser. Fiz.*, **28**, No. 11, pp. 6–29, 1985.
17. S.A. Akhmanov, M.A. Vorontsov, V.P. Kandidov, A.P. Sukhorukov, S.S. Chesnokov, "Optical beams thermal blooming and methods of its compensation," *Izv. V.U.Z. Radiofiz.*, **23**, No. 1, pp. 1–37, 1980.
18. J.A. Fleck, J.R. Morris, "Equivalent thin lens model for thermal blooming compensation," *Appl. Opt.*, **17**, No. 16, pp. 2575–2579, 1978.
19. F.P. Vasil'ev, M.A. Vorontsov, O.L. Litvinova, "About optimal controlling of the thermal blooming process," *J. Comput. Math. Math. Phys.*, **19**, No. 4, pp. 1053–1058, 1979.
20. O.I. Vasil'ev, S.S. Lebedev, "Optical beam thermal blooming correction in transition media by a WFC-mirror," *Atm. Opt.*, **3**, No. 2, pp. 174–181, 1990.
21. J. Herrmann, "Properties of phase conjugate COAT," *J. Opt. Soc. Am.*, **65**, p. 1212, 1975.
22. T.J. Karr, "Atmospheric effects on laser propagation," *Proc. SPIE Vol.* **1060**, pp. 120–128, 1989.
23. J.F. Schonfeld, "Linearized theory of thermal-blooming phase-compensation instability with realistic adaptive-optics geometry," *J. Opt. Soc. Am. B.*, **9**, No. 10, pp. 1803–1812, 1992.
24. V.P. Lukin, B.V. Fortes, "Influence of wavefront dislocations on instability phase conjugation of thermal blooming compensation," *Atm. Oceanic Opt.*, **8**, No. 3, pp. 435–447, 1995.
25. N.B. Baranova, B.J. Zel'dovich, "Dislocations of surface of wavefront and amplitude zeros," *JETF*, **80**, p. 1789, 1981.

CHAPTER 5

A Laser Reference Beacon as a Key Element of an Adaptive Optics System

In the preceding chapters on estimating the efficiency of adaptive systems, we analyzed models of a turbulent atmosphere and considered methods for solving the wave equation that describes propagation of optical radiation in the atmosphere. The next important problem is the development of a combined model of an adaptive system using a reference source as one of its key elements.

The importance of research into the efficiency of adaptive optics systems with an artificial reference source was understood by the late 1970s. In that period (and even earlier), the main principles of current adaptive electro-optics systems were formulated. According to these principles, the reference source is an element that helps to obtain information on the distribution of fluctuations in the propagation channel of optical radiation. The method of formation of the reference source affects the structure of the system as a whole. If the reciprocity principle is the basis for an adaptive system, then the most appropriate adaptive scheme is one with an independent reference source generating a beam propagating in direction opposite to the direction of the radiation to be corrected.

For practical implementation of the emitting optical system, the atmosphere should be included in the feedback loop; i.e., backward scattering should be taken into account, with radiation reflected from an object or inhomogeneities of the atmosphere. In this way, an artificial (virtual) reference source is formed.

In the early 1980s, artificial reference sources were named *laser guide stars* in adaptive astronomy. Such stars became possible through the use of Rayleigh backscattering or elastic scattering of atmospheric aerosols at the altitudes from 8 to 20 km or stimulated emission in clouds of atomic metals (for example, sodium). In the first case, a laser guide star is referred to as a Rayleigh guide star and in the second as a sodium guide star.

The requirements for a laser source that forms a Rayleigh star are loose. On the other hand, the requirements for the wavelength, bandwidth, and power of a laser source forming a sodium star are quite strict. This is due to the selectivity and saturation of absorption at stimulated emissions in sodium vapor at altitudes from 85 to 100 km.

5.1 Some Features of Reflected Optical Wave Fluctuations in a Turbulent Atmosphere

Interest in this problem arose because of the need to generate reference radiation as a result of reflection from some objects in the atmosphere (radiation of a beacon or a guide star). Light can be reflected in the atmosphere from inhomogeneities such as clouds or aerosols and from real objects. As an example, we consider the radiation specularly reflected from an object. In this case, we have to take into account the peculiarities of fluctuations of optical parameters of radiation passed twice through atmospheric inhomogeneities. In adaptive phase control algorithms, the most important parameter is the phase of the reflected wave, so special attention should be paid to phase fluctuations.

In the mid-1970s, scientists and engineers working with optical systems of viewing and beaming in the atmosphere as well as atmospheric sensing systems understood the importance of allowing for peculiarities of fluctuations of the reflected waves. In contrast to transmitting systems, optical sensing systems always involve the effect of double passage through the atmosphere. The sensing radiation passes through the same optical inhomogeneities two times: at forward and backward propagation.

5.1.1 Enhanced backscatter

For the situation when an object is illuminated and viewed through the same turbulent atmosphere, J.C. Dainty et al. and A.S. Gurvich, A.N. Bogaturov, and V. A. Myakinin coined the term double passage imaging. The physical basis of double passage imaging lies in the fact that reversible paths of illumination and viewing are coherent with one another, regardless of the phase distortion of the turbulence, and thus double passage imaging is closely linked to the phenomenon of enhanced backscatter encountered when dense volume media and highly sloped rough surfaces are illuminated [1].

The most complete overview of this problem is given in Refs. [2] and [3], which summarize the results of studying the effect of fluctuations of the atmospheric refractive index on optical waves under the conditions of double propagation through the same atmospheric region. This occurs as laser radiation is reflected from a target or waves are backscattered by atmospheric aerosols. In this case, the fluctuations of optical waves are determined by correlations between fluctuations in the incident and backscattered waves that have passed through the same inhomogeneities in the turbulent atmosphere. This may lead to qualitatively new properties of fluctuations compared with forward propagation, i.e., enhancement of backscattering of intensity fluctuations, long-range correlations in the field of the reflected wave, and so on.

The important role of Russia in this field was confirmed at the International Meeting for Wave Propagation in Random Media "Scintillation" held in Seattle,

Washington, in August 1992 [4]. The following Russian scientists have been working in this field: Yu. A. Kravtsov, A. N. Malakhov, A.S. Gurvich, K.S. Gochelashvily, V.I. Shishov, A.I. Saichev, V.A. Banakh, V.A. Mironov, V.U. Zavorotny, V.I. Klyatskin, A.I. Kon, V.I. Tatarskii, Yu.N. Barabanenkov, S.S. Kashkarov, G.Ya. Patrushev, V. P. Aksenov. M.I. Charnotskii, M.L. Belov, V.M. Orlov, I.G. Yakushkin, Z.I. Feizulin, A.G. Vinogradov, A. B. Krupnik, L. Apresyan, and others.

5.1.2 Phase fluctuations of specularly reflected waves

In the mid-1970s, Soviet scientists and particularly the scientists working with the Institute of Atmospheric Optics Siberia RAS were involved in an investigation of the peculiarities of phase fluctuations in sensing schemes employing mirror reflectors, i.e., fluctuations of optical waves that have passed through the atmosphere twice.

The problem of detection and ranging of mirror objects appeared in some practical applications, for example, as a part of a project using laser range finding with a corner retroreflector and in problems of atmospheric sensing, laser interferometry, wind velocity measurement by Doppler meters, and so on. In our opinion, the most interesting papers concerning this problem are the publications by Smith et al. [6, 7]. Following these papers, many other publications devoted to phase fluctuations of specularly reflected waves appeared in the period from 1974 to 1980.

At the Institute of Atmospheric Optics, experimental and theoretical investigations of the problem were begun in 1976 [8-18]. In Ref. [10], the increase (compared with a direct wave) of phase fluctuations for a wave specularly reflected from an object in a backward direction was demonstrated theoretically. Experimental studies carried out [11] in 1975–1976 confirmed the theoretical conclusions. Later (in 1977) the efficiency of compensation for phase distortions was considered [12]. The correction of phase fluctuations was assessed from the data characterizing the reflected wave for a plane mirror with a radius much larger than the beam radius used as a retroreflector. More detailed analysis of phase fluctuations in reflected waves was performed in 1980 [13]. In Ref. [13], the correlation and structure functions of phases for optical waves were calculated for a specularly reflected wave.

Special attention was paid to the phenomenon of *phase fluctuation coupling.* It was shown that some peculiarities are specific to statistical characteristics of the phase and the relevant characteristics for reflected waves [14-18]. These peculiarities should be allowed for in evaluations of optical radar signals as well as in adaptive optics systems employing phase conjugation algorithms. Several experiments in the real atmosphere with correction of random wavefront tilt were carried out in 1976–1980 [16-18].

5.1.3 Random displacements of the image of a sensed volume in a turbulent atmosphere

Fluctuations of an image centroid were considered by Orlov et al. [3]. The image was formed through an inhomogeneous medium in a sensing system with the use of a focused laser beam. In particular, fluctuations of the image of a sensed volume were studied. Monostatic and bistatic schemes were considered. Focused and collimated laser beams were used to irradiate the sensed volume. The equations describing the variance of centroid fluctuations have been derived for an image in the photodetector plane without any restrictions on the reflective properties of the object ([3], pp. 84–95).

It was shown that for strong dispersion on a surface (Lambertian approximation) in a bistatic sensing scheme, the variance of linear displacements of the image centroid can be written as ([3], p. 92)

$$\left\langle \rho_{\mathrm{im}}^2 \right\rangle = \frac{F^2}{x^2} \left\langle \rho_{\mathrm{lb}}^2 \right\rangle + F^2 \left\langle (\varphi_F^{\mathrm{ss}})^2 \right\rangle, \tag{5.1.1}$$

where $\left\langle \rho_{\mathrm{lb}}^2 \right\rangle$ is the variance of random displacements of the beam centroid in a sensing plane (it was assumed that the beam propagates upward) and $\left\langle (\vec{\rho}_F^{\,\mathrm{ss}})^2 \right\rangle$ is the variance of random angular displacements of the image of a "secondary" immobile source (downward propagation). Thus, it was shown that for a bistatic scheme, the variance of angular displacements of the image is a sum of angular displacements of the sensing beam and the image of the "secondary" immobile source. If a focused beam is used in a strongly scattering medium, then the secondary source is, in reality, a point.

In that period, calculations were performed for the cases when the "secondary" source was treated as a point (a small scattering volume) and as an object with a finite volume. For example, in the paper by Kallistratova and Kon [5], the jitter of the image of a thin luminous string was considered.

Thus, we can conclude that in the USSR in the early 1980s, scientists understood that under some conditions a volume can be considered as infinitely small (a point source or a laser guide star), but in other problems it is necessary to take into account the size of this volume; i.e., if an object is large enough, averaging over its volume is necessary, as was done in Ref. [5]. At the same time, Orlov et al. [3] failed to correctly calculate the correlation between fluctuations of focused beam displacements and displacements of the image of the reference (secondary) sources. This was done in 1979–1980 by Lukin [8, 9].

Let us calculate the correlation between random displacements of the energy centroid of an optical beam that has passed through a layer of a turbulent medium and the centroid of some image formed by the optical system through the same medium [8]. This can be an image of a reference source (beacon) or an optical beam reflected from a flat mirror with an infinite radius. Random displacements of the beam centroid are described by the vector [19]:

$$\vec{\rho}_c = \frac{1}{2P_0}\int_0^x d\xi(x-\xi)\iint d^2R I(\xi,\mathbf{R})\nabla_R \varepsilon_1(\xi,\mathbf{R}), \tag{5.1.2}$$

where $\varepsilon_1(\xi,\mathbf{R})$ are fluctuations of permittivity at the point $(\xi,\mathbf{R})$, $I(\xi,\mathbf{R})$ is the intensity [at the point $(\xi,\mathbf{R})$] of the field generated by the source located in the plane $\xi = 0$, and x is the thickness of the turbulent layer, $P_0 = \iint d^2RI(0,\mathbf{R})$.

The random displacements of the image in the focal plane of the optical system (equivalent to a thin lens with a focal length F and area $\Sigma = \pi R_0^2$) are described by the equation

$$\vec{\rho}_{cF} = -\frac{F}{k\Sigma}\iint_\Sigma \nabla_\rho S(x,\vec{\rho})d^2\rho, \tag{5.1.3}$$

where k is the radiation wave number and $S(x,\rho)$ are the fluctuations of the optical wave phase over the aperture of the optical system (in the plane $\xi = x$) at the point $\vec{\rho}$. The correlation between the random vectors $\vec{\rho}_c$ and $\vec{\rho}_{cF}$ can be represented as

$$K = \langle \vec{\rho}_c \vec{\rho}_{Fc}\rangle / \left[\langle(\vec{\rho}_c)^2\rangle\langle(\vec{\rho}_{cF})^2\rangle\right]^{1/2}. \tag{5.1.4}$$

Hereafter, $\langle ...\rangle$ denotes averaging over the ensemble of realizations of the random function $\varepsilon_1(\xi,\mathbf{R})$.

To make the situation clearer, let us consider the correlation between random displacements of the centroid of a Gaussian beam and the centroid of the image for an infinite plane wave (the beam and the plane wave propagate along the same optical path). In what follows we assume that the functions $< I(\xi,\mathbf{R}) >$ and $\Phi_\varepsilon(x\ ,\kappa)$ are isotropic and mean intensity is described by the following equation [19]:

$$\langle I(\xi,\mathbf{R})\rangle = \frac{a^2}{a_{\text{Eff}}^2(\xi)} e^{-R^2/a_{\text{Eff}}^2(\xi)}, \tag{5.1.5}$$

where

$$a_{\text{eff}}^2(x') = a^2\left\{\left(1-\frac{x}{f}\xi\right)^2 + \Omega^{-2} + \Omega^{-2}\left[\frac{1}{2}D_S(2a)\right]^{6/5}\right\},$$

$\Omega = ka^2/x$, a and f are the initial parameters of the Gaussian beam, and $D_S(2a)$ are phase structure functions. Thus we obtain

$$K = \int_0^1 d\xi (1-\xi) \int_0^\infty d\kappa \kappa^3 \Phi_n(\kappa) \exp\left[-\frac{\kappa^2 (R_0^2 + a_{\text{eff}}^2)}{4}\right] \cos\left[\frac{\kappa^2 x(1-\xi)}{2\kappa}\right] \Bigg/$$
$$\Bigg/ \left[\int_0^1 d\xi (1-\xi)^2 \int_0^\infty d\kappa \kappa^3 \Phi_n(\kappa) \exp\left(-\frac{\kappa^2 a_{\text{eff}}^2}{2}\right)\right]^{1/2} \times$$
$$\times \left[\int_0^1 d\xi \int_0^\infty d\kappa \kappa^3 \Phi_n(\kappa) \times \exp\left(-\frac{\kappa^2 R_0^2}{2}\right) \cos^2\left(\frac{\kappa^2 x(1-\xi)}{2\kappa}\right)\right]^{1/2}. \tag{5.1.6}$$

The form of K is similar for a spherical wave, as well when these waves or the Gaussian beam are reflected from a flat reflector with an infinite radius.

In calculations, we use the following spectrum:

$$\Phi_n(\kappa) = 0.033 C_n^2 (\kappa^2 + \kappa_0^2)^{-11/6}, \tag{5.1.7}$$

which accounts for deviation from the power series near the outer scale $L_0 = 2\pi\kappa_0^{-1}$.

As an example, let us estimate the correlation between displacements of the image for a plane wave at the focus of a telescope and random displacements of a beam with an initial diameter equal to the diameter of the telescope's entrance pupil. The estimation is performed for a homogeneous path; the parameters of the problem are the following:

$$\kappa_0^{-1} >> \max\left(R_0,\ a_{\text{eff}},\ \sqrt{x/k}\right),\ kR_0^2 >> x,\ \Omega^{-2}\left[\frac{1}{2} D_s(2a)\right]^{6/5} << 1.$$

The resulting value of K is approximately equal to 0.84. Thus, it was shown that there is a high positive correlation between the displacements of the Gaussian beam and the displacements of the plane wave centroid provided that the beam and the image-forming wave propagate on the same path and in the same direction.

Owing to the relatively high correlation, the algorithm for controlling the correction of random angular displacements of a beam $\vec{\rho}_c/x$ can be based on the expression $\alpha(a/2R_0)^{1/3} \rho_{cF}/F$, where α is the coefficient of the loop, which is chosen to ensure minimum residual angular displacements of the beam

$$\min\left\langle\left[\frac{\vec{\rho}_c}{x}-\frac{\alpha(a/2R_0)^{-1/3}\vec{\rho}_{cF}}{F}\right]^2\right\rangle.$$

Thus, a transition from linear measurements to angular ones makes it possible to control a laser beam's position by using the data measured on the image of the reference source [8].

These results [8] were later generalized (in 1980) for the case when the beam and the image-forming wave propagate in opposite directions [9]. It was assumed that the image is formed in the focal plane of a telescope for the following scenarios:

- plane wave, spherical wave, Gaussian beam, and
- radiation reflected from a plane mirror.

For a plane wave propagating over a homogeneous path and for a broad beam it was found that with a collimated beam K = −0.87, and for a focused beam it is −0.82. For spherical and any other waves, the results could be obtained from the data published in Ref. [9].

So, as early as 1979 the sign of the correlation was determined and its value was estimated. And, in Ref. [9] (1980) the fundamental possibility of using radiation backscattered by atmospheric aerosol was first mentioned. These problems are considered in greater detail in the Sec. 5.3. Summing up, we can conclude that during 1976–1980 Soviet scientists obtained all the functions necessary to analyze random displacements of the image of a sensed object for both bistatic and monostatic schemes.

However, in solving a particular problem, a model of the scattering (or reflecting) medium, which in turn determines the model of the secondary source [see Eq. (5.1.1)] is still an open question. In this case, two approaches are possible: direct introduction of a model of the source and solution of the problem of backscattering.

5.2 Improvement of the Quality of an Atmospheric Image by Adaptive Optics Methods

Consider now the possibilities of improving the quality of atmospheric images formed by an adaptive optics system. To correct the image of a star formed by a telescope, two approaches are largely used: one based on measurement of the field generated by the reference source and the other based on maximization of the sharpness function.

In this section we consider correction by an adaptive optics system functioning as a system with feedback [21] and utilizing information about the instantaneous distribution of inhomogeneities in the propagation medium. For this we use the algorithms of adaptive control based on the reciprocity [22] principle.

The quality of the image of an extended object formed by an optical system in a turbulent atmosphere can be improved with the help of adaptive correction. To do this, information about the distribution of turbulent inhomogeneities of the medium along the propagation path is extracted from the measured fields of the reference source [23,24]. As an object with a known amplitude-phase distribution, a reference source located at a known distance from the receiver can be formed directly on the surface of the object being imaged; it can be located at infinity (a reference star); or, finally, it can be located between the object and the objective [23].

Let the extended object be in the plane x_{obj}, the point reference source be in the plane x_{ref}, and the entrance aperture of the telescope be in the plane x_0. The distribution of the field U_{obj} of the extended object formed in the x_0 plane is described by the equation [20, 25]

$$U(x_0,\vec{\rho}) = \iint d^2\rho_1 U_{\text{obj}}(\vec{\rho}_1)\; G(x_0,\vec{\rho};\; x_{\text{obj}},\vec{\rho}_1)\,. \tag{5.2.1}$$

$G(x_0,\vec{\rho};\; x_{\text{obj}},\vec{\rho}_1)$ is the Green's function of the turbulent atmosphere between the planes x_0 and x_{obj}. The entrance aperture of the telescope is denoted as $W(\rho)$, and the action of the telescope is replaced by an equivalent lens introducing the phase term $\exp(-\imath k\rho^2/2f)$, where f is the equivalent focal length of the telescope.

We assume that the correction is based on the phase conjugation algorithm using the phase of the wave measured from the reference source. In this regard, if the reference source is so small or so far away that it is not resolved by the optical system of the telescope, it can be treated as a point source. The phase of the wave (with wave number $k = 2\pi/\lambda$) from the reference source in the entrance aperture plane x_0 can be written as

$$S_{\text{ref}}(x_0,\vec{\rho}) = \frac{k\rho^2}{2(x_{\text{ref}} - x_0)} + S(x_0,\vec{\rho};x_{\text{ref}},0)\,, \tag{5.2.2}$$

where $S(x_0,\vec{\rho};x_{\text{ref}},0)$ is the random phase that is due to turbulence of a spherical wave as it propagates from the plane x_{ref} to the plane x_0. We assume here that the point source is located on the optical axis of the telescope. This means that the conditions of the experiment allow us to form the reference source on the same optical axis as that on which the object is located. Here we omit discussing the techniques for formation of reference sources.

There are a great number of techniques for separating the diffraction phase $k\rho^2/2(x_{\text{ref}} - x_0)$ from $S(x_0, \vec{\rho}; x_{\text{ref},} 0)$ in Eq. (5.2.2). With the purely turbulent contribution to the phase of the reference source [Eq. (5.2.2)] thus available to us, we can solve the correction problem using either the phase conjugation algorithm or the total phase conjugation (TPC) algorithm.

The TPC algorithm uses the total phase via Eq. (5.2.2) to correct the wavefront and therefore forms a beam of arbitrary geometry only in the plane of the reference source. If we succeed in whatever way in separating the diffraction phase from the turbulent phase in Eq. (5.2.2), then we can apply the PC algorithm for beam formation in an arbitrary plane (which is different from the plane of the reference source) or for image correction. Here, however, it is necessary for the turbulence-induced phase of the reference source measured in the x_0 plane to coincide with the phase of the object field over the entire path to the object.

This can take place, for example, along slant paths, where the turbulent intensity $C_n^2(h)$ decreases with height h. The reference source upon which the experimental scheme in question is based should be located somewhat higher than the effective layer of atmospheric turbulence H_{eff}, and the choice of its location is governed by the condition

$$\frac{\int\limits_{H_0}^{\infty} dh\, C_n^2(h) - \int\limits_{H_0}^{H_{\text{eff}}} dh\, C_n^2(h)}{\int\limits_{H_0}^{\infty} dh\, C_n^2(h)} \leq \varepsilon,$$

where ε is determined by the admissible level of residual distortions in the optical wave in which the aberration-free image is being formed. Here it is assumed that the object is located practically at infinity, i.e., outside the atmosphere; H_0 is the height of the entrance aperture of the telescope.

Let us consider the phase conjugation algorithm in detail, employing it only within the aperture $W(\rho)$. Then the corrected field is written as

$$U_c(x_0, \vec{\rho}) = W(\rho) \frac{G^*(x_0, \vec{\rho};\ x_{\text{ref}}, 0)}{G_0^*(x_0, \vec{\rho};\ x_{\text{ref}}, 0)} \iint d^2\rho_1\ U_{\text{obj}}(\vec{\rho}_1)\ G(x_0, \vec{\rho};\ x_{\text{obj}}, \vec{\rho}_1), \tag{5.2.3}$$

where $G_0^*(x_0, \vec{\rho}; x_{\text{ref}}, 0)$ is a Green-function for free space propagation. The ratio $G^*(x_0, \vec{\rho}; x_{\text{ref}}, 0)/G_0^*(x_0, \vec{\rho}; x_{\text{ref}}, 0)$ is called the "correction function." The corrected field in the image plane of the object $x'_{\rm obj}$ is formed in the same way as in a "thin lens":

$$U_{\rm im}(x'_{\rm obj},\vec{\rho}) = \iint d^2\rho_2 U_c(x_0,\vec{\rho}_2)\exp(-ik\rho_2^2/2f)\; G(x'_{\rm obj},\,\vec{\rho};\,x_0,\,\vec{\rho}_2)\,. \tag{5.2.4}$$

Here it is assumed that the image is formed in a vacuum in the optical system of the telescope. Summing Eqs. (5.2.3) and (5.2.4), we have

$$\begin{aligned} U_{\rm im}(x'_{\rm obj},\vec{\rho}) &= \iint d^2\rho_1 d^2 r_2 W(\vec{\rho}_1)\exp(-ik\rho_1^2/2f)G_0(x'_{\rm obj},\vec{\rho};x_0,\vec{\rho}_1) \\ &\times U_{\rm obj}(\vec{r}_1)G^*(x_0,\vec{\rho}_1;x_{\rm ref},0)G(x_0,\vec{\rho}_1;x_{\rm obj},\vec{r}_1)/G_o^*(x_0,\vec{\rho}_1;x_{\rm ref},0). \end{aligned} \tag{5.2.5}$$

As a result, if the ratio

$$G^*(x_0,\vec{\rho}_1;x_{\rm ref},0)/G_0^*(x_0,\vec{\rho}_1;x_{\rm ref},0) = \exp\left[-iS(x_0,\vec{\rho}_1;x_{\rm ref},0)\right],$$

then the mean intensity distribution in the image plane x' is

$$\begin{aligned} \langle I_{\rm im}(x'_{\rm obj},\,\vec{\rho})\rangle &= \iint d^4\rho_{1,2} d^4 r_{1,2}\exp\left[-ik(\rho_1^2-\rho_2^2)/2f\right]W(\vec{\rho}_1) \\ &\times W(\vec{\rho}_2)G_0\left(x_0,\,\vec{\rho}_1;x_{\rm obj},\,\vec{\rho}\right)G_0^*\left(x_0,\,\vec{\rho}_1;x_{\rm obj},\,\vec{\rho}\right)U_{\rm obj}(\mathbf{r}_1)U_{\rm obj}^*(\mathbf{r}_2) \\ &\times G_0^*\left(x_0,\,\vec{\rho}_2;x_{\rm obj},\mathbf{r}_2\right)G_0\left(x_0,\,\vec{\rho}_1;x_{\rm obj},\mathbf{r}_1\right) \\ &\times\langle\exp\{i[S(x_0,\,\vec{\rho}_1;x_{\rm ref},0) \\ &-S(x_0,\,\vec{\rho}_2;x_{\rm ref},0)]-\left[S\left(x_0,\,\vec{\rho}_1;x_{\rm obj},\mathbf{r}_1\right)-S\left(x_0,\,\vec{\rho}_2;x_{\rm obj},\mathbf{r}_2\right)\right]\}\rangle. \end{aligned} \tag{5.2.6}$$

Here we have used the phase approximation [2] for Green's function expressed as $G = G_0\exp(iS)$, where S is the random phase caused by turbulence calculated in the geometric optics approximation and the angular brackets indicate averaging over the ensemble of turbulent fluctuations.

Below, for convenience we denote the expression in the angular brackets in Eq. (5.2.6) simply as $\langle\ldots\rangle$. In calculating $\langle\ldots\rangle$, we assume that the fluctuations of the phase S are Gaussian, from which it follows that

$$\langle ... \rangle = \exp\left[-\frac{1}{2} D_s(x_o, \vec{\rho}_1 - \vec{\rho}_2; x_{\text{ref}}, 0) - \frac{1}{2} D_s(x_o, \vec{\rho}_1 - \vec{\rho}_2; x_{\text{obj}}, \mathbf{r}_1 - \mathbf{r}_2) \right.$$
$$+ \langle S(x_0, \vec{\rho}_1; x_{\text{ref}}, 0) S(x_0, \vec{\rho}_1; x_{\text{obj}}, \mathbf{r}_1) \rangle$$
$$- \langle S(x_0, \vec{\rho}_2; x_{\text{ref}}, 0) S(x_0, \vec{\rho}_1; x_{\text{obj}}, \mathbf{r}_2) \rangle \qquad (5.2.7)$$
$$- \langle S(x_0, \vec{\rho}_1; x_{\text{ref}}, 0) S(x_0, \vec{\rho}_2; x_{\text{obj}}, \mathbf{r}_2) \rangle$$
$$\left. + \langle S(x_0, \vec{\rho}_2; x_{\text{ref}}, 0) S(x_0, \vec{\rho}_2; x_{\text{obj}}, \mathbf{r}_2) \rangle \right],$$

where $D_s(x_0, \vec{\rho}_1 - \vec{\rho}_2; x_{\text{obj}}, \mathbf{r}_1 - \mathbf{r}_2)$ is the phase structure function of two spherical waves whose sources are separated by the displacement $\mathbf{r}_1 - \mathbf{r}_2$ and observation points are separated by the vector $\vec{\rho}_1 - \vec{\rho}_2$. Let us consider the correlation functions of the type $\langle S(x_0, \vec{\rho}_1; x_{\text{ref}}, 0) S(x_0, \vec{\rho}_2; x_{\text{obj}}, \mathbf{r}_1) \rangle$ in more detail. Since the random phase in the geometric optics approximation can be represented in the form [20]

$$S(x_0, \rho; x_{\text{ref}}, 0) = k \int_{x_0}^{x_{\text{ref}}} d\xi \iint dn(\kappa, \xi) \exp\left[i\kappa\rho \frac{\xi}{(x_{\text{ref}} - x_0)} \right], \qquad (5.2.8)$$

the correlation can be written as

$$\langle S(x_{\text{ref}}, \rho_1) S(x_{\text{obj}}, \rho_2) \rangle = k^2 \int_{x_0}^{x_{\text{ref}}} d\xi_2 \iint \langle dn(\kappa_1, \xi_1) dn(\kappa_2, \xi_2) \rangle$$
$$\times \exp\left\{ i \left[\kappa_1 \rho_1 \frac{\xi_1}{(x_{\text{ref}} - x_0)} + \kappa_2 \rho_2 \frac{\xi_2}{(x_{\text{ref}} - x_0)} \right] \right\}. \qquad (5.2.9)$$

Using the representation [20]

$$\langle dn(\kappa_1, \xi_1) dn(\kappa_2, \xi_2) \rangle = 2\pi \delta(\xi_1 - \xi_2) \delta(\kappa_1 + \kappa_2) \Phi_n(\kappa_1, \xi_1) d^2\kappa_1 d^2\kappa_2,$$

we can transform Eq. (5.2.9) into the following form:

$$\langle S(x_{\text{ref}}, \vec{\rho}_1) S(x_{\text{obj}} \vec{\rho}_2) \rangle = 2\pi k^2 \int_{x_0}^{x_{\text{ref}} \le x_{\text{obj}}} d\xi \iint d^2\kappa \Phi_n(\kappa, \xi)$$
$$\times \exp\left\{ i\vec{\kappa}\xi / (x_{\text{ref}} - x_0) \left[\vec{\rho}_1 - \vec{\rho}_2 \left(\frac{x_{\text{ref}} - x_0}{x_{\text{obj}} - x_0} \right) \right] \right\},$$

where $\Phi_n(\kappa, \xi)$ is the spectral density of the refractive index fluctuations. Thus, in the geometric optics approximation,

$$\left\langle S(x_0,\vec{\rho}_1;x_{\text{ref}},0)S(x_0,\vec{\rho}_2;x_{\text{obj}},\mathbf{r}_1)\right\rangle = B_S\left[x_0,\ \vec{\rho}_1-\vec{\rho}_2;x_{\text{ref}},\mathbf{r}_1\left(\frac{x_{\text{ref}}-x_0}{x_{\text{obj}}-x_0}\right)\right]. \tag{5.2.10}$$

Now, taking Eq. (5.2.10) into account, we bring Eq. (5.2.7) into the form

$$\begin{aligned}\langle\ldots\rangle = \exp\Bigg\{&-\frac{1}{2}D_S(x_0,\rho_1-\rho_2;x_{\text{ref}},0)-\frac{1}{2}D_S(x_0,\rho_1-\rho_2;x_{\text{obj}},\mathbf{r}_1-\mathbf{r}_2)\\ &-\frac{1}{2}D_S\left[x_0,0;x_{\text{ref}},\mathbf{r}_1\left(\frac{x_{\text{ref}}-x_0}{x_{\text{obj}}-x_0}\right)\right]\\ &+\frac{1}{2}D_S\left[x_0,\rho_1-\rho_2;x_{\text{ref}},\mathbf{r}_1\left(\frac{x_{\text{ref}}-x_0}{x_{\text{obj}}-x_0}\right)\right]\\ &+\frac{1}{2}D_S\left[x_0,\rho_1-\rho_2;x_{\text{ref}},\mathbf{r}_2\left(\frac{x_{\text{ref}}-x_0}{x_{\text{obj}}-x_0}\right)\right]\\ &-\frac{1}{2}D_S\left[x_0,0;x_{\text{ref}},\mathbf{r}_2\left(\frac{x_{\text{ref}}-x_0}{x_{\text{obj}}-x_0}\right)\right]\Bigg\}.\end{aligned} \tag{5.2.11}$$

Using the representation of Green's function in free space, we can rewrite Eq. (5.2.6) as

$$\begin{aligned}\left\langle I_{im}(x'_{\text{obj}},\vec{\rho})\right\rangle = &\frac{k^4}{16\pi^2(x_0-x'_{\text{obj}})^2(x_{\text{obj}}-x_0)^2}\iint d^4\rho_{1,2}d^4r_{1,2}W(\rho_1)W^*(\rho_2)\\ &\times U_{\text{obj}}(\vec{r}_2)\exp\left[-ik\frac{(\rho_1^2-\rho_2^2)}{2f}+ik\frac{(\vec{\rho}_1-\vec{\rho})^2}{2(x_0-x'_{\text{obj}})}\right.\\ &\left.-ik\frac{(\vec{\rho}_2-\vec{\rho})^2}{2(x_0-x'_{\text{obj}})}+ik\frac{(\vec{r}_1-\vec{\rho}_1)^2}{(x_{\text{obj}}-x_0)}-ik\frac{(\vec{r}_2-\vec{\rho}_2)^2}{2(x_{\text{obj}}-x_0)}\right]\langle\ldots\rangle.\end{aligned} \tag{5.2.12}$$

Let us first consider the diffraction terms in Eq. (5.2.12), which control the formation of the image in a vacuum. We choose the plane x', which is conjugate to the plane x, as the image plane. From the condition for conjugate planes, we have

$$1/(x_0-x'_{\text{obj}})-1/(x_{\text{obj}}-x_0)=1/f\ , \tag{5.2.13}$$

and since $(x_{obj} - x_0) >> (f,\ x_0 - x'_{obj})$ we arrive at the conclusion that

$$1/(x_0 - x'_{obj}) = 1/f.$$

Consequently, remote objects are practically always imaged at the focus. Based on this result, it is possible to simplify the diffraction terms in Eq. (5.2.12). It follows that in a vacuum the intensity distribution in the focal plane has the following form:

$$I(f,\vec{\rho}) = \frac{k^4}{16\pi^2 f^2 (x_{obj} - x_0)^2} \iint d^4\rho_{1,2} d^4 r_{1,2} W(\rho_1) W^*(\rho_2) U_{obj}(\vec{r}_1)$$
$$\times U^*_{obj}(\vec{r}_2) \exp\left\{ -ik \frac{\vec{\rho}(\vec{\rho}_1 - \vec{\rho}_2)}{f} + ik \frac{\left[(\vec{r}_1 - \vec{\rho}_1)^2 - (\vec{r}_2 + \vec{\rho}_2)^2 \right]}{2(x_{obj} - x_0)} \right\}. \tag{5.2.14}$$

If we assume a Gaussian object $U_{obj}(r) = \exp(-r^2/2a^2_{obj})$ (here a_{obj} is the effective radius of the object), we have

$$I(f,\vec{\rho}) = \frac{k^2 \Omega^2_{obj}}{4\pi^2 f^2 (1 + \Omega^2_{obj})} \iint d^4\rho_{1,2} W(\rho_1) W^*(\rho_2) \exp\left\{ -ik \frac{\vec{\rho}(\vec{\rho}_1 - \vec{\rho}_2)}{f} \right.$$
$$\left. + i \frac{k}{2(x_{obj} - x_0)} \left[\rho_1^2 / (1 - i\Omega_{obj}) - \rho_2^2 / (1 + i\Omega_{obj}) \right] \right\}, \tag{5.2.15}$$

where $\Omega_{obj} = k a^2_{obj} / (x_{obj} - x_0)$.

Let us turn back to the case of image correction. Here it should be noted that there is one term (the second) in the exponent in Eq. (5.2.11) that determines fluctuations in the system without correction. This term depends on the integration over the entire path from the object to the telescope. The rest of the terms in Eq. (5.2.11) are associated with correction. In calculations we use the following equation for the phase structure function:

$$D_S(x, \vec{\rho}_1 - \vec{\rho}_2;\ x',\ \vec{r}_1 - \vec{r}_2) = 2.91 k^2 \int_{x'}^{x} d\xi C_n(\xi) \left| \frac{(\xi - x')}{(x - x')} (\vec{\rho}_1 - \vec{\rho}_2) \right.$$
$$\left. + \frac{(x - \xi)}{(x - x')} (\vec{r}_1 - \vec{r}_2) \right|^{5/3}. \tag{5.2.16}$$

The structure characteristic of the refractive index in the integrand in Eq. (5.2.16) depends on the integration variable C_n^2 along the propagation path.

Let us consider the case of vertical propagation (a telescope pointed at the zenith). In calculations we make allowance for the altitude dependence of C_n^2. This dependence is realized in practice in the form of various models.

Let us carry out an estimation based on the following model [20]:

$$C_n^2(x) = \begin{cases} C_{n_0}^2 (x/x_0)^{-2/3}, & x_0 \le x \le x_t, \\ C_{n_1}^2 (x/x_t)^{-4/3}, & x > x_t, \end{cases} \tag{5.2.17}$$

where $C_{n_0}^2$ is the value of the structure characteristic at the initial altitude, x_t is the altitude at which the two-thirds power dependence goes over to the four-thirds power dependence, and $C_{n_1}^2 = C_{n_0}^2 (x_1/x_0)^{-2/3}$. Thus, this model is determined by the parameters x_t and $C_{n_0}^2$. In calculations connected with Eq. (5.2.11), we require that $\min(x_{obj}, x_{ref}) > x_t$. Substituting Eq. (5.2.16) into Eq. (5.2.11), after combining like terms, we have

$$\begin{aligned} \langle \ldots \rangle = \exp \Bigg\{ & -1.46k^2 |\vec{\rho}_1 - \vec{\rho}_2|^{5/3} (x_{obj} - x_0)^{-5/3} \int_{x_0}^{x_{o\ddagger}} (\xi - x_0)^{5/3} C_n^2(\xi)\, d\xi \\ & -1.46k^2 |\vec{\rho}_1 - \vec{\rho}_2|^{5/3} (x_{obj} - x_0)^{-5/3} \int_{x_{o\bullet}}^{x_{o\ddagger}} (x_{obj} - \xi)^{5/3} C_n^2(\xi)\, d\xi \\ & -1.46k^2 |\vec{\rho}_1 - \vec{\rho}_2|^{5/3} \int_{x_0}^{x_{o\bullet}} \left[\left(\frac{x_{obj} - \xi}{x_{obj} - x_0} \right)^{5/3} - \left(\frac{x_{ref} - \xi}{x_{ref} - x_0} \right)^{5/3} \right] C_n^2(\xi)\, d\xi \Bigg\}. \end{aligned} \tag{5.2.18}$$

Here, note that $\vec{r}_1$ and $\vec{r}_2$ are the variables of integration over the object, and $\vec{\rho}_1$ and $\vec{\rho}_2$ are the variables of integration over the aperture of the telescope. Analysis of Eq. (5.2.18) reveals that the term containing $|\vec{r}_1 - \vec{r}_2|^{5/3}$ (the object variables) remains the same as in the absence of correction. This is the term that is determined by the action of so-called anisoplanatism of the atmosphere.

This parameter is connected [23, 24] with the problem of constructing the image of an extended object. There are two terms in Eq. (5.2.18) containing $|\vec{\rho}_1 - \vec{\rho}_2|^{5/3}$ (the variables of the entrance aperture of the telescope), one of which is determined by inhomogeneities in the region between the reference source and the object (this part remains uncompensated since the reference source is located closer to the objective than the object) and another is determined by the inhomogeneities between the objective and the reference source (its presence is due to the inequality of the curvature of the spherical waves arriving from the reference source and from the object's plane).

It is clear that for the model of Eq. (5.2.17) the altitude region corresponding to the two-thirds power law gives the largest contribution and therefore it is always necessary to choose $x_{\mathrm{ref}} > x_t$. In this regard, all the integrals in Eq. (5.2.18) can be divided into two parts:

$$\int_{x_0}^{x_{\mathrm{obj}}} d\xi(\ldots) = \int_{x_0}^{x_t} d\xi(\ldots) + \int_{x_t}^{x_{\mathrm{obj}}} d\xi(\ldots). \tag{5.2.19}$$

If the region corresponding to the two-thirds power law is absent from the spectrum in Eq. (5.2.17), then the first term in Eq. (5.2.19) is equal to zero. Substituting the model of Eq. (5.2.17) into Eq. (5.2.18), we arrive at the equation

$$\begin{aligned}\langle\ldots\rangle = \exp\Bigg\{ &-1.46k^2|\vec{r}_1 - \vec{r}_2|^{5/3}\Bigg[\frac{11}{6}C_{n_0}^2 \frac{x_0^{2/3}(x_t - x_0)^2}{x_{\mathrm{obj}}^{5/3}} \\ &+\frac{3}{4}C_{n_1}^2\left(\frac{x_t^{4/3}}{x_0^{1/3}} - \frac{x_t^{4/3}(x_t - x_0)^{4/3}}{x_{\mathrm{obj}}^{5/3}}\right)\Bigg] - 1.46k^2|\vec{\rho}_1 - \vec{\rho}_2|^{5/3}\frac{5}{4}C_{n_0}^2 x_0^{2/3} \\ &\times\left(\frac{x_t^{4/3}}{x_{\mathrm{ref}}} - \frac{x_0^{4/3}}{x_{\mathrm{ref}}}\right) - 1.46k^2|\vec{\rho}_1 - \vec{\rho}_2|^{5/3}\,3C_{n_1}^2\Big[x_t^{4/3}\left(x_t^{-1/3} - x_{\mathrm{ref}}^{-1/3}\right) \\ &- x_t\left(1 - x_t/x_{\mathrm{ref}}\right)^{5/3}\Big]\Bigg\}.\end{aligned} \tag{5.2.20}$$

Now let us consider each of the terms in Eq. (5.2.20) separately. First, let us turn our attention to the term containing $|\vec{r}_1 - \vec{r}_2|^{5/3}$. It determines what objects can be seen clearly and in their entirety in the objective at a given level of turbulence along the path and do not undergo any change upon correction. Consequently, correction with one reference source does not remove the problem of anisoplanatism, and for correction of the image of a large object it is necessary to form several reference sources. For further calculation, we introduce the parameter r_a, the radius of the isoplanar region, as follows:

$$\begin{aligned}1.46k^2|\vec{r}_1 - \vec{r}_2|^{5/3}\Bigg\{&\frac{11}{6}C_{n_0}^2\frac{x_0^{2/3}(x_t - x_0)^2}{x_{\mathrm{obj}}^{5/3}} \\ &+\frac{3}{4}C_{n_1}^2\left[x_t^{4/3}/x_{\mathrm{obj}}^{1/3} - x_t^{4/3}(x_t - x_0)^{4/3}/x_{\mathrm{obj}}^{5/3}\right]\Bigg\} = |\vec{r}_1 - \vec{r}_2|^{5/3}/r_a^{5/3}.\end{aligned} \tag{5.2.21}$$

In this context, we can say that the object is located in the isoplanar region if $a_{\text{obj}} < r_a$. The parameter r_a has a simple physical meaning. From Eq. (5.2.21), taking into account only the first (and main) term, we have

$$r_a = \frac{x_{\text{obj}}}{(x_t - x_0)} \left[2.68 k^2 C_{n_0}^2 x_0^{2/3} (x_t - x_0)^{1/3} \right]^{-3/5},$$

or

$$\frac{r_a}{x_{\text{obj}}} = \frac{\left[2.68 k^2 C_{n_0}^2 x_0^{2/3} (x_t - x_0)^{1/3} \right]^{-3/5}}{(x_t - x_0)}.$$

Thus, it turns out that the region of isoplanatism in the object's plane is discernible from a distance that is equal to the range of the object at the same angle at which the coherence length of the atmospheric layer is discernible through this layer.

Let us consider the term containing $|\vec{\rho}_1 - \vec{\rho}_2|^{5/3}$:

$$1.46 k^2 |\vec{\rho}_1 - \vec{\rho}_2|^{5/3} \left\{ \frac{5}{4} C_{n_0}^2 x_0^{2/3} \left(x_t^{4/3} - x_0^{4/3} \right) / x_{\text{ref}} + C_{n_1}^2 \left[x_t \times \left(1 - x_t^{1/3} / x_{\text{ref}}^{1/3} \right) - x_t (1 - x_t / x_{\text{ref}})^{5/3} \right] \right\}. \quad (5.2.22)$$

Note that for $x_t >> x_0$ this term reduces to

$$1.825 \times k^2 |\vec{\rho}_1 - \vec{\rho}_2|^{5/3} C_{n_0}^2 x_0^{2/3} x_t^{4/3} / x_{\text{ref}}.$$

Compare it with the same term in the case of no correction:

$$10.22 \times k^2 |\vec{\rho}_1 - \vec{\rho}_2|^{5/3} C_{n_0}^2 x_0^{2/3} x_t^{1/3}.$$

The action of correction is characterized by a relative decrease in fluctuations that is proportional to x_t / x_{ref}. The term in Eq. (5.2.22) can be represented, in analogy with the phase structure function in the absence of correction, as $|\vec{\rho}_1 - \vec{\rho}_2|^{5/3} / r_{\text{eff}}^{5/3}$, where r_{eff} is the *effective coherence length.*

Summing up all these results, we obtain the following equation for an object having a Gaussian shape:

$$\langle I(f,\vec{\rho})\rangle = \frac{k^4}{16\pi^4 f^2 (x_{\text{obj}} - x_0)^2} \iint d^4\rho_{1,2} W(\rho_1) W^*(\rho_2)$$
$$\times \exp\left[-ik \frac{\vec{\rho}(\vec{\rho}_1 - \vec{\rho}_2)}{f} + ik \frac{(\rho_1^2 - \rho_2^2)}{2(x_{\text{obj}} - x_0)} \right] \iint d^4 r_{1,2} \exp\left[-\frac{r_1^2}{2a_{\text{obj}}^2} \right.$$
$$\times (1 - i\Omega_{\text{obj}}) - ik \frac{\vec{r}_1 \vec{\rho}_1}{(x_{\text{obj}} - x_0)} - \frac{r_2^2}{2a_{\text{obj}}^2} (1 + i\Omega_{\text{obj}}) \tag{5.2.23}$$
$$\left. + ik \frac{\vec{r}_2 \vec{\rho}_2}{(x_{\text{obj}} - x_0)} - |\vec{r}_1 - \vec{r}_2|^{5/3} / r_a^{5/3} - |\vec{\rho}_1 - \vec{\rho}_2|^{5/3} / r_{\text{eff}}^{5/3} \right],$$

which can be compared with the diffraction-limited result of Eq. (5.2.12). Calculating the diffraction integral gives the following results:

$$\iint d^2 r_1 (\ldots) = 2\pi a_{\text{obj}}^2 / (1 - i\Omega_{\text{obj}}) \exp\{ -k\rho_1^2 \Omega_{\text{obj}} / [2(x_{\text{obj}} - x_0)(1 - i\Omega_{\text{obj}})] \}$$

for a circular aperture

$$W(\rho) = \begin{cases} 1, & \rho \le R, \\ 0, & \rho > R, \end{cases}$$

obtained by

$$\iint d^2\rho_1 W(\rho_1) \exp\left\{ -\rho_1^2 \left[\frac{k}{2(x_{\text{obj}} - x_0)} \times \frac{\Omega_{\text{obj}}}{(1 - i\Omega_{\text{obj}})} - i \frac{k}{2(x_{\text{obj}} - x_0)} \right] \right.$$
$$\left. -ik \frac{\vec{\rho}\vec{\rho}_1}{f} \right\} = \pi R^2 \frac{J_1\left(\frac{k\rho R}{f}\right)}{\left(\frac{k\rho R}{2f}\right)} - \pi R^2 \frac{\Omega_R}{4} \left(\frac{\Omega_{\text{obj}} - i}{1 + \Omega_{\text{obj}}^2} \right)$$
$$\times {}_1F_2\left(2; 1, 3; \frac{k^2 \rho^2 R^2}{4f^2} \right); \quad \Omega_R = \frac{kR^2}{(x_{\text{obj}} - x_0)}.$$

Finally we arrive at the following equation for $\Omega_R < \Omega_{\text{obj}}$:

$$I_0(\rho)=\frac{k^2R^4\Omega_{\text{obj}}^2}{\pi^2 f^2\left(1+\Omega_{\text{obj}}^2\right)}\left[\frac{J_1^2\left(\frac{k\rho R}{f}\right)}{\left(\frac{k\rho R}{f}\right)^2}-\frac{\Omega_R\Omega_{\text{obj}}}{4\left(1+\Omega_{\text{obj}}^2\right)}\times\frac{J_1\left(\frac{k\rho R}{f}\right)}{\left(\frac{k\rho R}{f}\right)}\right. \tag{5.2.24}$$

$$\left.\times {}_1F_2\left(2;1,3;\frac{k^2\rho^2R^2}{4f^2}\right)+\frac{\Omega_R^2}{64\left(1+\Omega_{\text{obj}}^2\right)}{}_1F_2^2\left(2;1,3;\frac{k^2\rho^2R^2}{4f^2}\right)\right]$$

when the image of a Gaussian object is formed in the telescope in the absence of turbulence.

To carry out the corresponding calculations in Eq. (5.2.23) without a computer, we use the quadratic approximation. Introduce the optical transfer function $\langle\tau(\vec{r})\rangle$ for an atmosphere–telescope system as [25]

$$\langle I(f,\vec{\rho})\rangle=\frac{k^2}{4\pi^2 f^2}\iint d^2r\langle\tau(\vec{r})\rangle\exp(-ik\vec{r}\vec{\rho}/f). \tag{5.2.25}$$

After some simple calculations we arrive at the equation

$$\langle\tau(\vec{r})\rangle=\frac{\Omega_{\text{obj}}^2}{\left(1+\Omega_{\text{obj}}^2+4\Omega_{\text{obj}}/\Omega_a\right)}\exp\left[-\frac{r^2}{r_{\text{eff}}^2}-\frac{kr^2}{4\left(x_{\text{obj}}-x_0\right)}\right.$$

$$\left.\times\frac{\Omega_{\text{obj}}\left(1+4\Omega_{\text{obj}}/\Omega_a\right)}{\left(1+4\Omega_{\text{obj}}/\Omega_a+\Omega_{\text{obj}}^2\right)}\right]\iint d^2xW(|\vec{x}+\vec{r}/2|)W(|\vec{x}-\vec{r}/2|)$$

$$\times\exp\left[-\frac{kx^2}{\left(x_{\text{obj}}-x_0\right)}\times\frac{\Omega_{\text{obj}}}{\left(1+4\Omega_{\text{obj}}/\Omega_a+\Omega_{\text{obj}}^2\right)}+ik\frac{\vec{x}\vec{r}}{4\left(x_{\text{obj}}-x_0\right)}\right.$$

$$\left.\times\frac{\left(1+4\Omega_{\text{obj}}/\Omega_a\right)}{\left(1+4\Omega_{\text{obj}}/\Omega_a+\Omega_{\text{obj}}^2\right)}\right],\quad \Omega_a=kr_a^2/\left(x_{\text{obj}}-x_0\right).$$

Calculating further the optical transfer function $\langle\tau(\vec{r})\rangle$ for a Gaussian aperture $W(x)=\exp\left(-x^2/2R^2\right)$, we obtain

$$\langle\tau(\vec{r})\rangle=\frac{\pi R^2\Omega_{obj}^2}{\left(1+4\Omega_{obj}/\Omega_a+\Omega_R\Omega_{obj}+\Omega_{obj}^2\right)}\exp\left\{-\frac{r^2}{4R^2}\left[1+4\frac{\Omega_R}{\Omega_{eff}}+\frac{\Omega_R\Omega_{obj}\left(1+4\Omega_{obj}/\Omega_a\right)}{\left(1+4\Omega_{obj}/\Omega_a+\Omega_{obj}^2\right)}+\right.\right.$$
$$\left.\left.+\frac{\Omega_R^2\left(1+4\Omega_{obj}/\Omega_a\right)^2}{\left(1+4\Omega_{obj}/\Omega_a+\Omega_{obj}^2\right)\left(1+4\Omega_{obj}/\Omega_a+\Omega_{obj}^2+\Omega_R\Omega_{obj}\right)}\right]\right\}. \qquad (5.2.26)$$

If the beam propagates through vacuum, we have the following equation:

$$\tau_0(r)=\frac{\pi R^2\Omega_{obj}^2}{\left(1+\Omega_R\Omega_{obj}+\Omega_{obj}^2\right)}\exp\left\{-\frac{r^2}{4R^2}\left[1+\frac{\Omega_R\Omega_{obj}}{\left(1+\Omega_{obj}^2\right)}+\frac{\Omega_R^2}{\left(1+\Omega_{obj}^2\right)+\left(1+\Omega_R\Omega_{obj}+\Omega_{obj}^2\right)}\right]\right\} \qquad (5.2.27)$$

for the optical transfer function of a telescope imaging the Gaussian beam.

Let us analyze the case $\left(\Omega_{obj}^2\rangle\rangle\Omega_{obj}/\Omega_a,\ \Omega_{obj}\rangle\rangle 1, \Omega_{obj}>\Omega_R\right)$ in detail. From Eqs. (5.2.26) and (5.2.27) we have

$$\langle\tau_0(r)\rangle=\tau_0(r)\exp\left(-\frac{r^2}{4R^2}\left\{\frac{4\Omega_R}{\Omega_{eff}}+\frac{4\Omega_R}{\Omega_a}+\Omega_R^2\left[\frac{\left(1+4\Omega_{obj}/\Omega_a\right)^2}{\Omega_{obj}^4}-\Omega_{obj}^{-4}\right]\right\}\right). \qquad (5.2.28)$$

Thus, the optical transfer function in Eq. (5.2.28) depends substantially on the ratio of the entrance radius of the telescope to the effective coherence length $\left(\Omega_R/\Omega_{eff}=R^2/r_{eff}^2\right)$ and the radius of the object to the radius of the isoplanatism zone $\left(\Omega_{obj}/\Omega_a=a_{obj}^2/r_a^2\right)$, as well as on combinations of these parameters. If the conditions $\Omega_R<\Omega_{eff}$, $\Omega_{obj}<\Omega_a$, and $\Omega_R<\Omega_{obj}$ are fulfilled, i.e., if the object occupies one isoplanar zone, then the effective coherence length exceeds the radius of the entrance aperture and the falloff scale of $<\tau(\vec{r})>$ coincides with the falloff scale of $\tau_0(r)$. It is easy to see that for a plane wave $(\Omega_{obj}\to\infty)$ the Fourier transform of Eq. (5.2.27), analogous to Eq. (5.2.25), is written as

$$I_0(\rho)=\frac{k^2R^4}{f^2}\exp\left(-k^2\rho^2R^2/f^2\right). \tag{5.2.29}$$

Both for the Gaussian and circular aperture [Eq. (5.2.26)] in a vacuum, the intensity distribution of the image decreases substantially at the distance $\rho \approx f/kR$. Correspondingly, the scale of the optical transfer function $\tau_0(r)$ is equal to the radius of the entrance aperture of the telescope R. Transforming Eq. (5.2.28), we have

$$\begin{aligned}\langle\tau(r)\rangle &= \tau_0(r)\exp\left\{-\frac{r^2}{4R^2}\left[4\frac{\Omega_R}{\Omega_{\text{eff}}}+4\frac{\Omega_R}{\Omega_a}+8\frac{\Omega_R^2}{\Omega_{\text{obj}}^2}\left(\frac{2}{\Omega_{\text{obj}}\Omega_a}+\frac{4}{\Omega_{\text{obj}}^2}\right)\right]\right\},\\ &\approx \tau_0(r)\exp\left[-r^2\left(1/r_{\text{eff}}^2+1/r_a^2\right)\right],\end{aligned} \tag{5.2.30}$$

where

$$\begin{aligned}\tau_0(r) &\cong \pi R^2\exp\left[-\frac{r^2}{4R^2}\left(1+\frac{\Omega_R}{\Omega_{\text{obj}}}+\frac{\Omega_R^2}{\Omega_{\text{obj}}^4}\right)\right]\\ &\approx \pi R^2\exp\left[-\frac{r^2}{4R^2}\left(1+\Omega_R/\Omega_{\text{obj}}\right)\right].\end{aligned} \tag{5.2.31}$$

Renormalize $\tau_0(r)$, as in Ref. [23], for $\tau_0(0)=1$. Employing the optical transfer function introduced in this way, we can calculate one of the functions that determine the quality of the combined atmosphere–telescope optical system, namely, the resolution [25, 21, 23]:

$$\Re=\iint d^2\kappa\,\langle\tau(\kappa)\rangle, \tag{5.2.32}$$

where $\kappa = kr/f$ is the spatial frequency. Integrating over the spatial frequencies directly in Eq. (5.2.26), we obtain

$$\begin{aligned}\Re &= \frac{2\pi k^2\Omega_{\text{obj}}^2R^2f^{-2}}{\left(1+4\Omega_{\text{obj}}/\Omega_a+\Omega_{\text{obj}}^2+\Omega_R\Omega_{\text{obj}}\right)}\left\{1+4\frac{\Omega_R}{\Omega_{\text{eff}}}+\frac{\Omega_R\Omega_{\text{obj}}\left(1+4\Omega_{\text{obj}}/\Omega_a\right)}{\left(1+4\Omega_{\text{obj}}/\Omega_a+\Omega_{\text{obj}}^2\right)}\right.\\ &\quad+\Omega_R^2\left(1+4\Omega_{\text{obj}}/\Omega_a\right)^2/\left[\left(1+4\Omega_{\text{obj}}/\Omega_a+\Omega_{\text{obj}}^2\right)\right.\\ &\quad\left.\left.\times\left(1+4\Omega_{\text{obj}}/\Omega_a+\Omega_R\Omega_{\text{obj}}+\Omega_{\text{obj}}^2\right)\right]\right\}.\end{aligned} \tag{5.2.33}$$

The resolution $\Re$, as a measure of the optical quality of the system, determines the minimal resolvable distance $\delta l \approx \left(1/2\sqrt{\Re}\right)$. The optical system has its maximum resolution in a vacuum and is determined by the parameters R, λ, and f for a planar wavefront:

$$\delta l_0 = \frac{f}{2\sqrt{2\pi k R}}.$$

In a turbulent medium, the limiting resolution (the minimum value of δl) for an arbitrarily large telescope (the limit $R \to \infty$) is determined by the coherence length r_0:

$$\delta l_{\min} \approx \frac{f}{2\sqrt{2\pi k r_0}}, \quad r_0 << R. \tag{5.2.34}$$

This circumstance lowers the efficiency of large telescopes. Practically speaking, $R = 2r_0$ represents the limiting resolution for large telescopes. The application of adaptive correction to a telescope increases its limiting resolution. Applying Eq. (5.2.33) to the casc $\Omega_{\text{obj}} >> \Omega_R$, $\Omega_{\text{obj}} >> 1$, we obtain

$$\Re \simeq 2\pi k^2 \Big/ \left[f^2 \left(\frac{1}{R^2} + \frac{4}{r_{\text{eff}}^2} + \frac{4}{r_a^2} \right) \right]. \tag{5.2.35}$$

Consequently, the resolution is determined by the minimal value of the telescope radius R, the effective coherence length r_{eff}, and the radius of isoplanatism r_a. Given the telescope radius and the height of the object (which determines the isoplanatism angle and the isoplanatism radius r_a), it is possible, by an appropriate choice of the location of the reference source x_{ref}, to increase r_{ref}. If we place the reference source on the object itself (in practice this means $r_{\text{eff}} = \infty$), we have

$$\Re \simeq 2\pi k^2 \Big/ \left[f^2 \left(\frac{1}{R^2} + \frac{4}{r_a^2} \right) \right]. \tag{5.2.36}$$

In this case the resolution of the system in the object's plane is limited by the radius of the zone of isoplanatism r_a. From Eq. (5.2.35) we have

$$\delta l = \frac{f}{2\sqrt{2\pi k R}} \left(1 + \frac{4R^2}{r_{\text{eff}}^2} + \frac{4R^2}{r_a^2} \right)^{1/2},$$

i.e., the objective of the telescope becomes diffraction limited when $R < r_a$, $R < r_{\text{eff}}$. If one of these conditions is not fulfilled, the quality of the optical system decreases. In this case, the radius of the objective having the limiting efficiency is

$$R \geq 2 \Big/ \left[\frac{1}{r_{\text{eff}}^2} + \frac{1}{r_a^2} \right]^{1/2}, \tag{5.2.37}$$

and for the reference source on the object $(r_{\text{eff}} = \infty)$,

$$R_{\text{lim}} \geq 2r_a. \tag{5.2.38}$$

Thus, we find that by choosing the position of the reference source, we can change the radius of the aperture that gives the maximum efficiency [Eq. (5.2.37)] all the way to the limit R_{lim} [Eq. (5.2.38)].

In conclusion, we can state that an adaptive optics system operating in the atmosphere and using a reference source can substantially increase its efficiency. The height of the reference source is chosen according to both the form of the C_n^2 profile along the propagation path and the level of permissible residual distortions. The optical system forms an aberration-free image of only one isoplanar region. If the angular dimensions of the object exceed this region, then it is not possible to obtain a diffraction-limited image using only one reference source. The size of this isoplanar region depends linearly on the distance, and the isoplanatism angle (the ratio of the object's radius to the distance from the observation plane to the object) coincides with the angle at which the coherence length of the atmosphere is discernible within the effective layer from a distance equal to the thickness of this layer. By making the appropriate choice of the location of the reference source, the efficiency of the optical device from the point of view of its resolution can be brought up to its limiting level determined by the radius of the isoplanar region; consequently, starting from heights for which $r_a > R$, we have the diffraction-limited resolution [21, 23].

Owing to reciprocity [22] (of fluctuations), these results extend to the problem of focusing an optical beam through the atmosphere with the help of a reference source. If the problem is to focus optical radiation in the plane of the reference source, the total phase conjugation algorithm can be used by invoking the entire phase of Eq. (5.2.2), but only if the initial beam is collimated. An interesting situation arises when the wavelengths of the beam formed and the reference source are different (let λ_1 be the wavelength of the reference beam and λ_2 be that of the beam formed through the atmosphere). Then the plane of optimal focusing (for an initial collimated beam) x_{foc} is related to the position of the reference source plane as follows:

$$\frac{k_1}{x_{\text{ref}} - x_0} = \frac{k_2}{x_{\text{foc}} - x_0}.$$

In order to focus the beam in the plane of the reference source, additional focusing is necessary with the curvature length

$$f_{\text{add}} = \frac{\lambda_1 (x_{\text{ref}} - x_0)}{\lambda_1 - \lambda_2}.$$

5.3 A Modern Concept of Adaptive Optics Systems with a Laser Guide Star

From 1982 to 1985, interest in the development of optical schemes employing artificial reference sources increased, as evidenced by the large number of papers published during this period. The studies in the field of adaptive optics were connected with developing a technique using the signal backscattered from atmospheric inhomogeneities to create a reference wave. It was noted that the signal backscattered from atmospheric inhomogeneities can be used to correct an image formed in ground-based astronomical telescopes.

The development of ground-based telescopes using the signal of a laser guide star became one of the most promising investigations in modern astronomy. The importance of research into the efficiency of artificial reference sources was understood as early as the late 1970s [21-23, 31]. The basic principles of the modern concept of adaptive optics systems were formulated during that period.

According to this concept, the reference source is a key element of the electro-optics scheme that is used to obtain information on the distribution of inhomogeneities in the medium's refractive index in the propagation channel. The structure of the electro-optics scheme as a whole depends significantly on the way in which the reference wave is formed. If the adaptive scheme is based on the reciprocity principle, then a scheme using an independent source of radiation propagating in the direction opposite to the direction of the beam to be corrected is most efficient. In practice, an adaptive optics system should include the atmosphere as a part of the closed feedback loop, i.e., account for the radiation backscattered from objects or atmospheric inhomogeneities. Artificial (virtual) reference sources such as laser guide stars are formed in this way.

The most promising method of improving a ground-based adaptive telescope is to equip it with an additional optical system for formation of a laser guide star [26-28]. A rather complete bibliography (unfortunately, not including the papers published by scientists from the USSR and the Russian Federation) on the main stages of development of the systems for the formation of laser guide stars is given in the paper by Ragazzoni [28].

This book is a logical continuation of work done from 1966 to 1986 in the USSR [2, 3, 5, 8, 9, 15, 19, 20, 21-23, 29-31], as well as the 1990s [32-36]. Theoretical and applied studies on the use of lasers for the formation of artificial reference stars have become popular in recent years, and this has forced us to return to some earlier results, although they were obtained about 20 years ago. Schemes for image correction using scattered laser radiation (see Fig. 5.3.1) were proposed in 1979–1986 in the USSR [8, 9, 23, 24]. These results were presented in detail in Refs. [8], [9], and [21], but became available to a wide range of investigators once a monograph on adaptive optics [24] was published (the fourth and fifth chapters of Ref. [24] are devoted to this problem). Many problems that are discussed now were solved during those years.

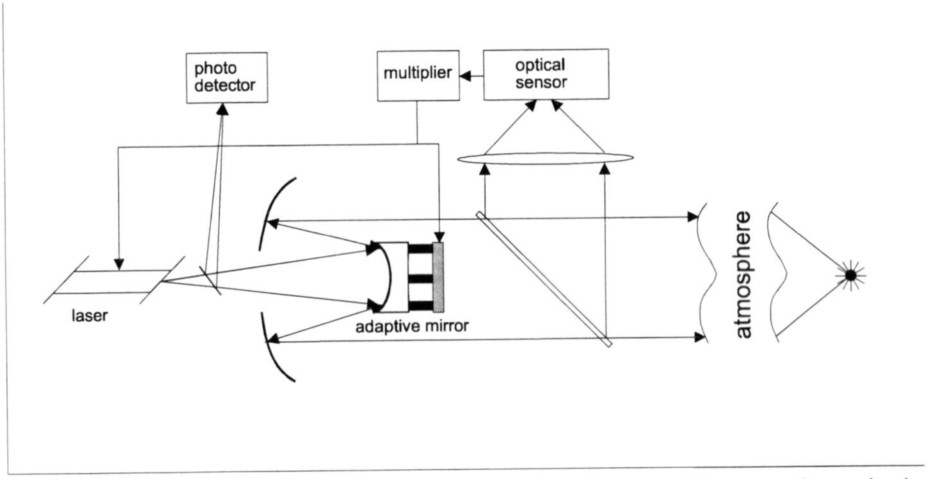

Figure 5.3.1. First approach to formation of a laser guide star through the atmosphere (1980). The optical system includes a laser, an adaptive mirror with a main telescope mirror, an additional lens, an optical sensor–optical measurer, an electric multiplier for adaptive mirror control, and a photodetector.

5.3.1 Some features of fluctuations of reflected waves

During the 1980s, the main principles were published that provide the operational basis of systems using reference sources to obtain information about fluctuations in the propagation channel of optical radiation. From a practical point of view, systems that use radiation backscattered from atmospheric inhomogeneities or an object are the most realizable. In this case, an artificial (virtual) reference source such as a laser guide star is formed [9]. There are two schemes for forming an LGS: monostatic and bistatic. The laser used is ground based and therefore the optical radiation passes twice through the same inhomogeneities, the first time at upward propagation to form the LGS and the second time at downward propagation as a result of backscattering (secondary emission, elastic aerosols, or molecular scattering) from atmospheric inhomogeneities. As indicated earlier, in both schemes, the peculiarities of fluctuations of optical waves that have passed through the atmosphere twice should be taken into account.

A number of terms have been coined to describe sensing schemes that account for this double passage through the atmosphere. They include *effective scattering volume*, a *monostatic optical scheme*, *bistatic laser sensing*, and some others.

5.4 Monostatic and Bistatic Schemes for Formation of a Laser Guide Star

In a monostatic scheme it is assumed that the laser beam is formed in the atmosphere in such a way that fluctuations for a laser guide star at forward and backward passage are maximally correlated. The contrary situation is characteristic of a bistatic scheme, in which it is assumed that there is no correlation between fluctuations caused by passage along the forward and backward paths.

Here we present results calculated for a "general" scheme of LGS formation with an arbitrary value for the correlation between random angular image displacements caused by fluctuations at the forward and backward paths. The equations for the monostatic and bistatic schemes in this case are derived as limiting cases.

While interest in this problem was aroused in connection with using a laser guide star signal to correct the image in a ground-based telescope, several serious difficulties arise, as was shown in particular by Fugate [27]. They are the influence of focal isoplanatism and the practical impossibility (for a monostatic scheme) of separating the contributions of upward and backward propagation to jitter of the LGS image.

In recent years papers have been published [37-39] that propose several approaches to solving one of these problems. Thus, in Refs. [38] and [39] the scheme of LGS formation uses a laser beam that passes through the main telescope; two auxiliary telescopes that measure LGS image jitter are also used for measurements. The optical scenario is such that the LGS is a pointlike object for the main telescope, but an extended object for the auxiliary telescopes. Therefore, as was shown in Ref. [38], the wavefront tilt correction in the main telescope cannot be used, but the bistatic scheme (with auxiliary telescopes) allows separation of the component in the LGS image jitter that corresponds to the directed laser beam and is highly correlated with the wavefront tilt for a natural star.

5.4.1 Correlation between random displacements of a laser beam and a natural star image for a bistatic scheme

Following the Ragazzoni approach [28], let us consider, as in Refs. [34] and [36], the following scheme for LGS formation (Fig. 5.4.1): The LGS is created by an auxiliary laser with an individual aperture. The parameters of the problem are the following: R_0 is the aperture radius of the main telescope; X is the altitude of (distance from) the LGS; the entrance pupil of the telescope is placed in the plane $\xi = 0$; a is the radius of the emitting aperture of the laser source; and $\vec{\rho}_0$ is the vector of displacement of the laser source with respect to the telescope optical axis. We presume that the tip-tilt measurements of the wavefront are conducted in the telescope with the use of an LGS formed at an altitude X above the entrance aperture exactly at the telescope's optical axis. The telescope is pointed toward the zenith, and a weak natural star and the LGS are both at the telescope's axis (or within the isoplanatic region out of the axis). The zenith angle of the laser beam is $|\vec{\rho}_0|/X$ (on the assumption that $|\vec{\rho}_0| << X$).

Let us also assume that the observed star (as a scientific object) has a plane wavefront. The vector characterizing a random tilt of this wavefront (function $\vec{\varphi}_F^{\,\mathrm{pl}}$) that is due to atmospheric turbulence is (see Ref. [20]) described by Eq.(5.1.3), where

$$S^{\mathrm{pl}}(0,\vec{\rho}) = k\int_0^\infty d\xi \iint d^2 n(\vec{\kappa}, x-\xi)\exp(i\vec{\kappa}\vec{\rho}) \qquad (5.4.1)$$

are phase fluctuations for a plane wave on the telescope's aperture and the following spectral expansion [19, 9] is used for fluctuations of the atmospheric refractive index:

$$n_1(\xi,\vec{\rho}) = \iint d^2 n(\vec{\kappa}, x-\xi)\exp(\iota\vec{\kappa}\vec{\rho}) .$$

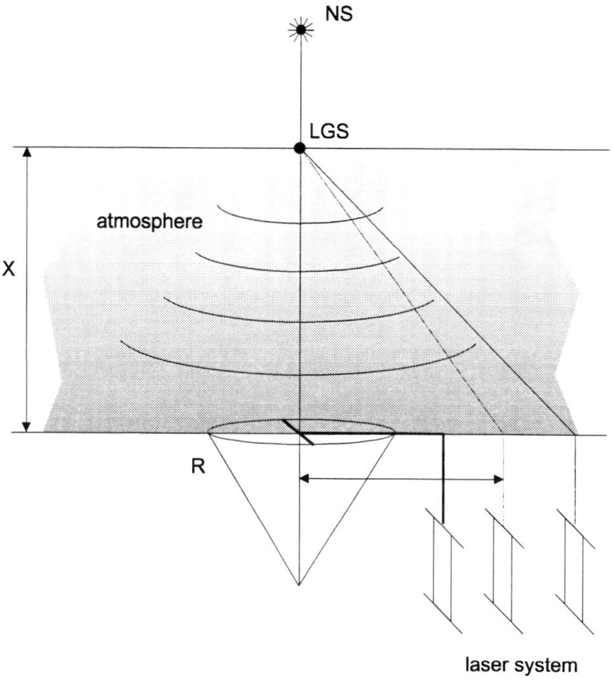

Figure 5.4.1. Scheme of formation of a laser guide star.

In Eq. (5.4.1) it is taken into account that the wave from the natural star passes downward. Random angular shifts of the centroid of the LGS [function $\vec{\varphi}_{\mathrm{lb}}(\vec{\rho}_0)$] formed by the laser source at the altitude X can be written [29] using Eq. (5.1.2) with

$$I = I\left[\xi, \vec{R} + \vec{\rho}_0(1 - \xi / X)\right].$$

In this equation, it is taken into account that the laser beam (see Fig. 5.3.1) is shifted by the vector $\vec{\rho}_0$ and its optical axis is tilted by the angle $|\vec{\rho}_0| / X$ with respect to the zenith.

Assume that the laser beam to be focused is sufficiently wide (so that $\Omega^{-1} = ka_0^2 / X >> 1$) and its extra broadening that is due to turbulence is insignificant. As viewed from the entrance pupil of the main telescope, the laser guide star can be considered a point source. At backward propagation, the additional angular jitter of such a spherical wave can be written [20] in the following form:

$$\vec{\varphi}_F^{\mathrm{sp}} = -\frac{1}{\Sigma}\iint_{\Sigma} d^2\rho \nabla_\rho S^{\mathrm{sp}}(0, \vec{\rho}) \tag{5.4.2}$$

to obtain the correlation function between random angular shifts of the natural star image [function $\vec{\varphi}_F^{\rm pl}$, Eq. (5.1.3)] formed by the telescope and shifts of the centroid of a focused beam formed by a tilted laser system [function $\vec{\varphi}_{\rm lb}(\vec{\rho}_0)$]. This formulation is based on our results published in Refs. [8], [9], [34], and [36]. The deviation of the turbulence spectrum from the power function in the region of large scales was accounted for [40-43] as

$$\Phi_n(\kappa,\xi) = 0.033 C_n^2(\xi)\kappa^{-11/3}\left[1-\exp(-\kappa^2/\kappa_0^2)\right], \qquad (5.4.3)$$

where $C_n^2(\xi)$ is turbulence intensity on the propagation path and $\kappa_0^{-1}(\xi)$ is the outer scale of turbulence. With allowance made for these properties, it is possible to obtain [8, 9, 34, 36] the following equation for the correlation function:

$$\begin{aligned}\left\langle \vec{\varphi}_{\rm lb}(\vec{\rho}_0)\vec{\varphi}_F^{\rm pl}\right\rangle &= \left[-2\pi^2 0.033\Gamma\left(\frac{1}{6}\right)\right]2^{1/3}R_0^{-1/3}\int_0^X d\xi C_n^2(\xi)(1-\xi/X)\\ &\left([1+b^2(1-\xi/X)^2]^{-1/6}\,{}_1F_1\left\{\frac{1}{6},1;-\frac{d^2(1-\xi/X)^2}{\left[1+b^2(1-\xi/X)^2\right]}\right\}\right.\\ &\left.-[1+b^2(1-\xi/X)^2+4c^2]^{-1/6}\,{}_1F_1\left\{\frac{1}{6},1;-\frac{d^2(1-\xi/X)^2}{\left[1+b^2(1-\xi/X)^2+4c^2\right]}\right\}\right).\end{aligned}$$

(5.4.4)

Here, the following designations are used: parameters $b = a/R_0$, $d = |\vec{\rho}_0|/R_0$, $c = \kappa_0^{-1}R_0^{-1}$, and ${}_1F_1(\ldots)$ is the Gaussian confluent hypergeometric function.

It can be readily shown that the second term of (5.4.4) in the braces is related to the outer scale of turbulence. For an infinite outer scale, this term can be neglected, so the correlation function takes the form:

$$\begin{aligned}\left\langle \vec{\varphi}_{\rm lb}(\vec{\rho}_0)\vec{\varphi}_F^{\rm pl}\right\rangle &= \left[-2\pi^2 0.033\ \Gamma\left(\frac{1}{6}\right)\right]2^{1/3}R_0^{-1/3}\int_0^x d\xi C_n^2(\xi)(1-\xi/X)\\ &\times\left[1+b^2(1-\xi/X)^2\right]^{-1/6}\,{}_1F_1\left\{\frac{1}{6},\ 1;-\frac{d^2(1-\xi/X)^2}{\left[1+b^2(1-\xi/X)^2\right]}\right\}.\end{aligned} \qquad (5.4.5)$$

The value $d = 0$ corresponds to a monostatic scheme of LGS. For the bistatic scheme $(d > 0)$, the condition $d >> 1$ corresponds to the asymptotic of a hypergeometric function ${}_1F_1(\ldots)$. Thus,

$$\left\langle \vec{\varphi}_{\text{lb}}(\vec{\rho}_0)\vec{\varphi}_F^{\text{pl}} \right\rangle = -2\pi^2 0.033\ \Gamma\left(\frac{1}{6}\right)2^{1/3}\Gamma^{-1}\left(\frac{5}{6}\right)R_0^{-1/3}d^{-1/3} \times \int_0^X d\xi C_n^2(\xi)(1-\xi/X)^{2/3}. \qquad (5.4.6)$$

From an analysis of Eq. (5.4.6) we can conclude that the correlation between the plane wave and the beam decreases to 0.1 for $d \geq 10^3$. This value is characteristic for a bistatic scheme (there is no correlation between the upward and downward propagation) for the infinite outer scale of turbulence.

Numerous experimentally obtained data [44–47] justify the assumption that the outer scale of turbulence $\kappa_0^{-1}(\xi)$ in the atmosphere is finite. Moreover, as a result of numerical estimations performed with different altitude profiles of $C_n^2(\xi)$ and $\kappa_0^{-1}(\xi)$, scientists have come to the conclusion that it is possible to introduce some characteristics (a spatial outer scale of coherence [48] or an effective outer scale of turbulence [49]) for a description of the entire atmospheric column. As it turned out, under moderate conditions of observation [50], the value of this effective outer scale is 5 to 60 m [49]. So for a telescope with R_0=4 m, the parameter $c = \kappa_0^{-1}R_0^{-1} < 10$.

Let us perform an asymptotic analysis of the influence of the outer scale on the correlation function expressed by Eq. (5.4.4). We take the variable d as an argument of the function and the variables b, c, and X as parameters. Simple estimations show that when the parameter $d = 0$ and $c < 5$, the correlation at the finite outer scale is 2 to 3 times lower than in the case of the infinite outer scale. With increasing d (for $d > 1$), the correlation calculated by Eq. (5.4.4) is no higher than 0.2. For $d > 2c$, the correlation is 17 times lower than that for the infinite outer scale. And finally, for $d >> c$, the sign of the correlation function $\left\langle \vec{\varphi}_{\text{lb}}(d)\ \vec{\varphi}_F^{\text{pl}} \right\rangle$ alternates and the dependence on d arises $\approx d^{-7/3}$.

To confirm this conclusion, let us numerically estimate the correlation coefficient

$$K(d,b,c,X) = \frac{\left\langle \vec{\varphi}_{\text{lb}}(\vec{\rho}_0)\vec{\varphi}_F^{\text{pl}} \right\rangle}{\sqrt{\left\langle [\vec{\varphi}_{\text{lb}}(\vec{\rho}_0)]^2 \right\rangle \left\langle (\vec{\varphi}_F^{\text{pl}})^2 \right\rangle}}. \qquad (5.4.7)$$

The coefficient can be expressed through the correlation function (5.3.4) and the corresponding variances:

$$\left\langle (\vec{\varphi}_F^{\text{pl}})^2 \right\rangle = \left[2\pi^2 0.033\ \Gamma\left(\frac{1}{6}\right)\right]2^{1/6}R_0^{-1/3}\int_0^x d\xi C_n^2(\xi)\left[1-\left(1+4c^2\right)^{-1/6}\right], \qquad (5.4.8)$$

$$\left\langle (\vec{\varphi}_{\text{lb}}(\vec{\rho}_0)^2\right\rangle = \left[2\pi^2 0.033\ \Gamma\left(\frac{1}{6}\right)\right] 2^{1/6} R_0^{-1/3} \int_0^x d\xi C_n^2(\xi)$$
$$\times\left\{\left[b^2(1-\xi/X)^2\right]^{-1/6} - \left[b^2(1-\xi/X)^2 + 4c^2\right]^{-1/6}\right\}. \tag{5.4.9}$$

Calculations have been performed with the model of $C_n^2(\xi)$ corresponding to moderate conditions of observation [50]. The altitudes corresponding to Rayleigh and sodium artificial stars (10 and 100 km) were included in the initial conditions. The parameter b was chosen as follows: b = 0.1, 0.3, 0.7, 1.0, 3.0, and 5.0. The values of b greater than unity are typical of a situation when a large telescope forms an artificial star for a smaller one. Such ratios of parameters can be obtained in observatories equipped with telescopes of different sizes, for example, at the Mauna Kea observatory, where the 10-m Keck telescope is used as an auxiliary one for a small telescope. In calculations, the values of c were the following: 1, 3, 5, 10, 100, and 1000. The case that c = 1000 in practice corresponds to the infinite outer scale.

The results obtained are shown (omitting the sign) in Figs. 5.4.2 and 5.4.3. All parameters are shown in the figures. It should be noted that the numerically obtained results confirm the conclusions of the above analytical analysis, so, it is possible to draw the following conclusions:

1. With large outer scales (c =100 and 1000), the scheme for LGS formation can be considered as bistatic (the upward and downward paths are uncorrelated) if the separation between the axes of the main and auxiliary telescopes is (200–1000) R_0; i.e., if $d > 200$.
2. With a finite outer scale ($c < 5$), separations of d that are 2 to 3 times larger than the outer scale indicate that the monostatic scheme should be replaced by the bistatic one (in which the correlation between upward and downward propagation is negligible).
3. Smaller values of d correspond to intermediate schemes (with partial correlation between the upward and downward paths).
4. It should be also noted that the results of our calculations do not coincide with the results reported in Refs. [39] and [53], since those papers considered the correlation functions for two plane waves traveling from infinity at different angles. In particular, in our calculations the correlation coefficient K for $d = 0$ is not equal to unity (with the minus sign). Only with decreasing X does the coefficient K asymptotically approach unity.
5. It is interesting to consider the behavior of K in the region of small values of c (c = 1, 3, 5) and relatively large values of b (b =3, 5). This situation is characteristic when a large telescope generates an artificial star for a smaller one. The results obtained (Figs. 5.4.2 and 5.4.3) show that in this scheme the correlation in the region $d < c$ is practically constant (equal to 0.4, 0.5, and 0.7, respectively).

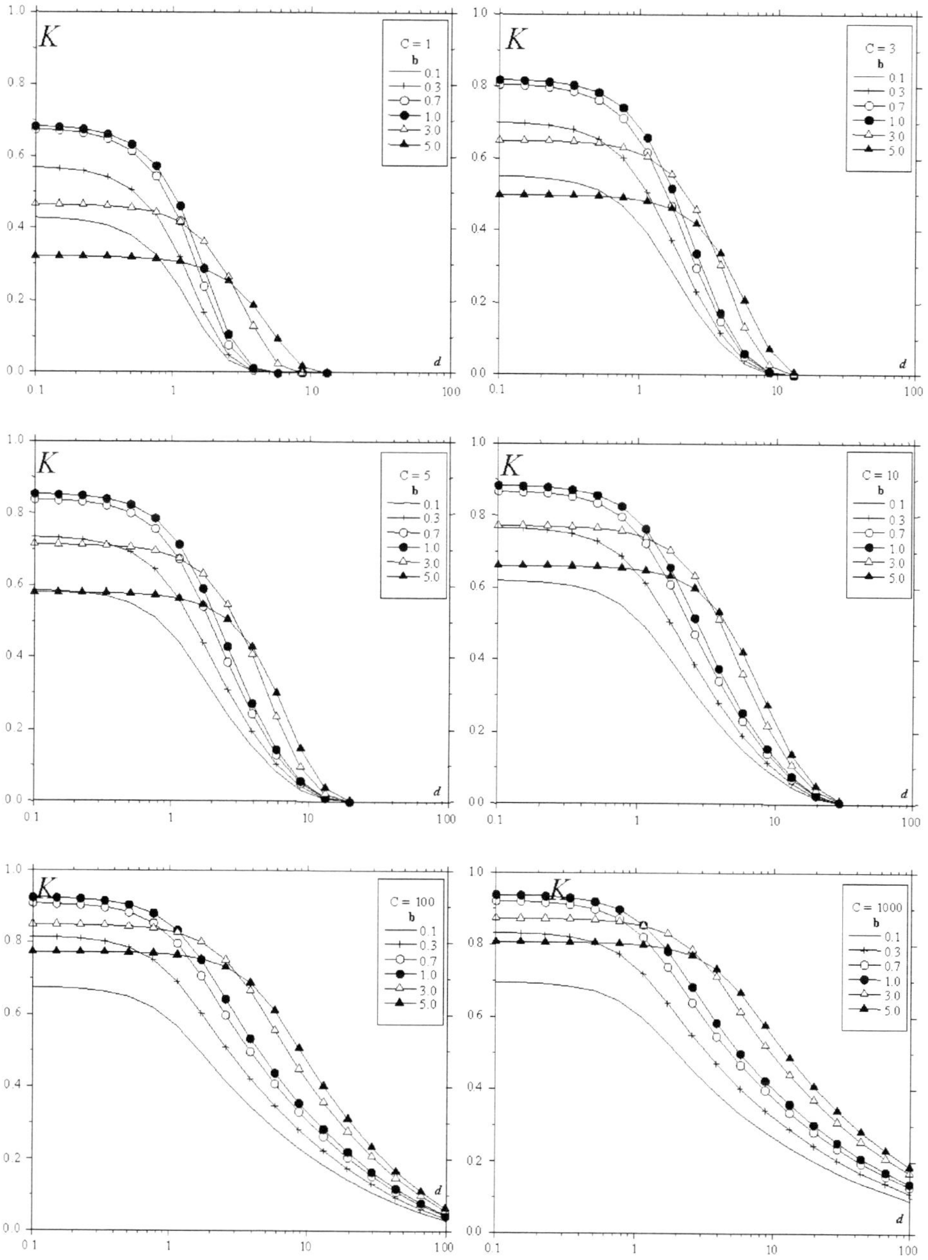

Figure 5.4.2. Correlation coefficient *K* from Eq. (5.4.7) for different values of the outer scale of turbulence (parameter *c*) and laser system aperture radius (parameter *b*); beacon altitude *X*=10 km.

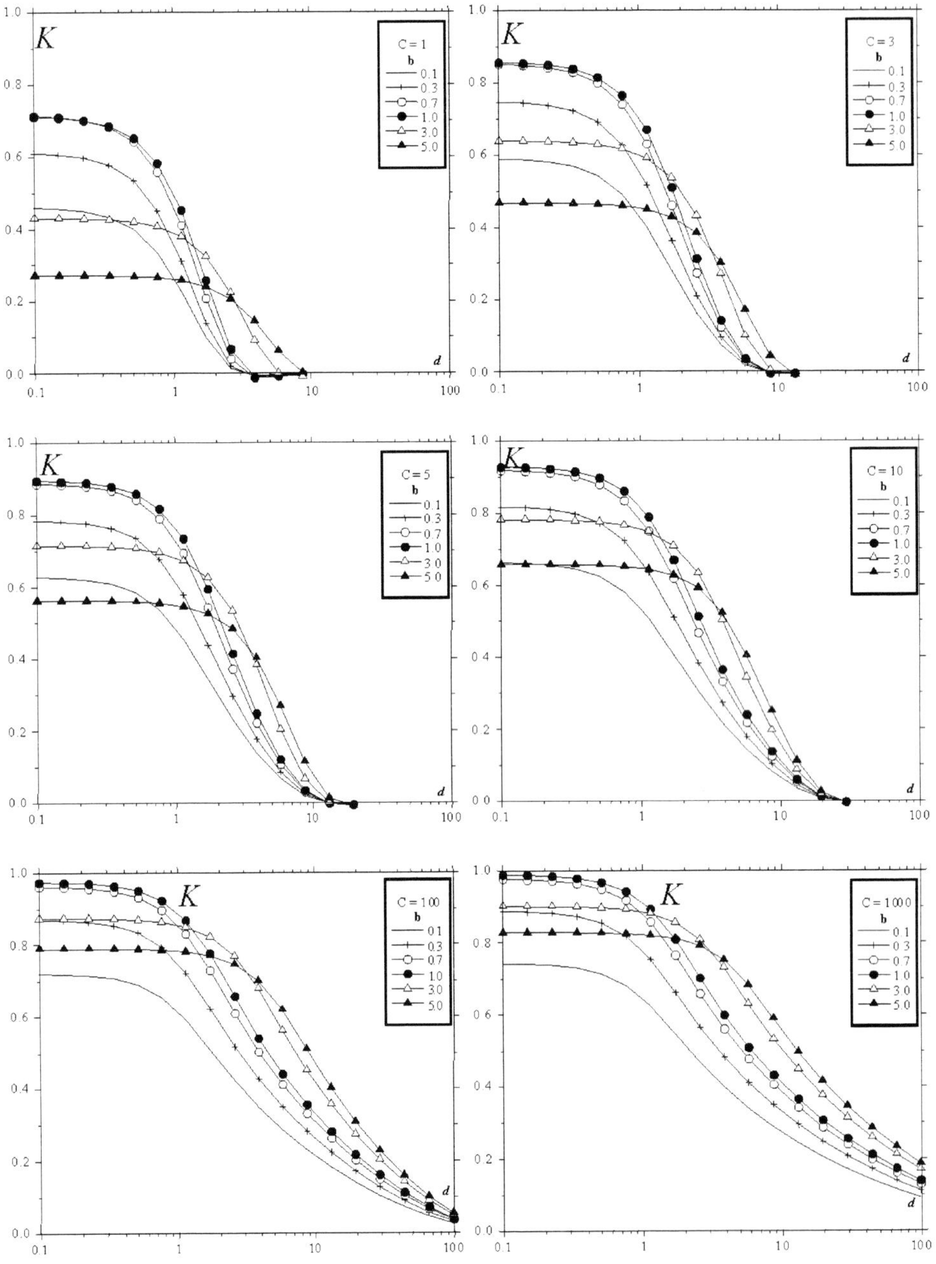

Figure 5.4.3. Correlation coefficient *K* from Eq. (5.4.7) for different values of the outer scale of turbulence (parameter *c*) and laser system aperture radius (parameter *b*); beacon altitude *X*=100 km.

6. With increasing c, the characteristic scale of the correlation function K increases; i.e., the correlation radius increases. This effect was mentioned in Ref. [39] (for two plane waves).
7. The increase in the correlation radius is not infinite; gradual saturation is observed in our numerical experiments. For small values of c (the outer scale is small), the correlation function $K(d)$ decreases to 0.1 for $d = c$. But even for $c = 100$, the correlation falls to 0.1 at $d = c/2$. For $c = 1000$, this decrease occurs at $d = c/10$. This takes place for the altitudes X=10 and 100 km.

The alternation of the sign of the correlation function predicted on the basis of asymptotic analysis [by comparing Eqs. (5.4.4) and (5.4.5)] is due to the finite size of the outer scale of turbulence. When the outer scale is small ($c = 1, 2, 3$) and $d > (2–3)c$, the correlation function alternates its sign. For large c, this effect is not observed. When the outer scale is large, the coefficient K [Eq. (5.4.7)] keeps its sign [51, 52].

It is interesting to find a relationship between the correlation length for two plane waves [53, 39] and the correlation length for a plane wave and a slanted laser beam (the latter is shown in Figs. 5.4.2 and 5.4.3). If determined, these data allow one to use the results of direct astronomical observations of the images of two stars viewed at different angles to predict the correlation (for tip-tilt correction) in the system of a telescope and a LGS, as well as to draw more correct conclusions on the mode of operation of this system. To our regret, it is impossible to simply compare the curves in Figs. 5.4.2 and 5.4.3 and the data presented in Ref. [39] because different models of atmospheric turbulence were used in the calculations. However, based on Figs. 5.4.2 and 5.4.3 [and Eqs. (5.4.4) and (5.4.5)], it is possible to estimate the real level of correlation between motion of the natural star image measured in the telescope and the auxiliary tilted laser beam that forms the LGS on the telescope's optical axis, and to obtain the parameters for calculating the efficiency of tip-tilt correction with any interesting parameters of the telescope, the size of the laser beam, the outer scale of turbulence, and the separation between the axes of the telescope and the laser beam.

The real level of correlation between the motion of the natural star image and the auxiliary tilted laser beam forming the LGS was calculated using a model spectrum of atmospheric turbulence. The structure parameters and outer scale of turbulence are the model's parameters. In this way, we have studied the peculiarities of the correlation connected with the finite outer scale of turbulence. Moreover, since all these parameters depend not only on the propagation path but also on the actual altitude of the telescope above sea level, possible variations of the model should be seriously discussed in relation to this aspect (model parameters), depending on the aerography of the underlying surface. Recall that numerous experimentally obtained data [44-47] justify the assumption that the outer scale $\kappa_0^{-1}(\xi)$ is finite. Moreover, as a result of numerical estimations performed with different altitude profiles of $C_n^2(\xi)$ and $\kappa_0^{-1}(\xi)$, scientists have come to the conclusion that it is possible to introduce some characteristics (spatial outer scale of coherence [48] or effective outer scale of turbulence [49]) to describe the entire atmospheric column.

At the same time, models of the spectrum of atmospheric turbulence that take into account the finiteness of the outer scale of turbulence and especially the varying value of the parameter $\kappa_0^{-1}(\xi)$ for vertical paths are rarely used. For homogeneous surface paths, it is shown that the value of κ_0^{-1} is fully finite and commensurable with the height above the underlying surface. However, for astronomical observations, a number of researchers consider the value of κ_0^{-1} to range from hundreds of meters to some kilometers [40]. In many observations, the results corresponded to an outer scale on the order of 1 m [44, 46].

Undoubtedly, the outer scale of the turbulence varies significantly both in the surface atmospheric layer and at high altitudes. Therefore, we cannot speak about the outer scale as having a definite value over the entire atmosphere. We propose to consider a number of possible versions of variations of the outer scale with the altitude h:

(A) $$\kappa_0^{-1}(h) = 0.4h,$$

(B) $$\kappa_0^{-1}(h) = \begin{cases} 0.4h & h \le 25\text{ m} \\ 2\sqrt{h}, & h > 25\text{ m} \end{cases}$$

(C) $$\kappa_0^{-1}(h) = \begin{cases} 0.4h, & h \le 25\text{ m} \\ 2\sqrt{h}, & 25\text{m} < h \le 2000\text{ m} \\ 89.4\text{ m}, & h > 2000\text{ m} \end{cases}$$

(D) $$\kappa_0^{-1}(h) = \frac{5}{\{1 + [(h - 7500)/2000]^2\}}$$

(E) $$\kappa_0^{-1}(h) = \frac{5}{\{1 + [(h - 7500)/2000]^2\}}.$$

Model (A) was recommended in Ref. [20] for use at low altitudes; model (B) is proposed by Fried in Ref. [25]; and model (C) is a generalization of the first two models. Models (D) and (E) are obtained as a generalization of the results of direct measurements in the United States, France, and Chile [25, 54-56]. Similar values of these parameters were obtained for the Mauna Kea Observatory (Hawaii) [25, 56]. Some investigators have cast doubt on these models [25, 44];

however, altitude variations in the outer scale within wide limits have gained recognition.

Figures 5.4.4 and 5.4.5 show the results calculated for the same characteristics as in Figs. 5.4.2 and 5.4.3, but for the outer scale as given by models (A–E) instead of the different values of the outer scale. The curves in Figs. 5.4.2 and 5.4.4 (as well as in Figs. 5.4.3 and 5.4.5) are very similar, but for models (D) and (E) the correlation level is lower for small separation between the axes of the main telescope and the laser beam.

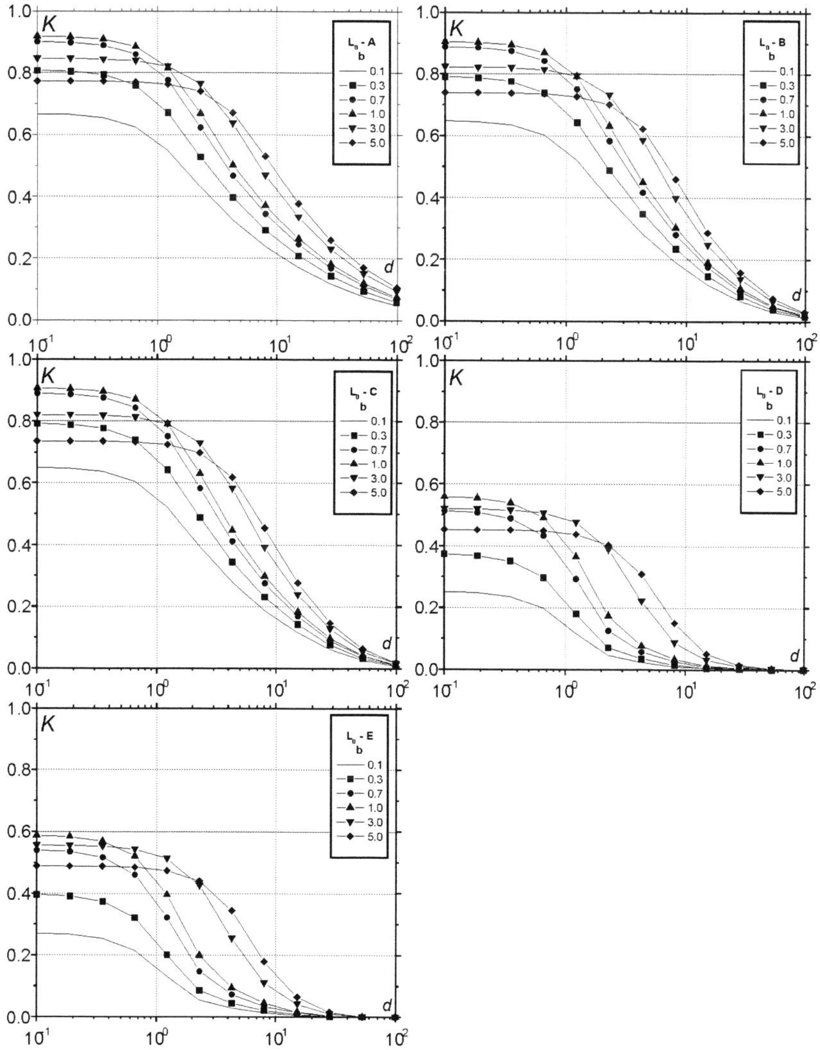

Figure 5.4.4. Correlation coefficient K from Eq. (5.4.7) for different models (A, B, C, D, E) of the outer scale of turbulence and the aperture radius of the laser system (parameter b); beacon altitude X=10 km.

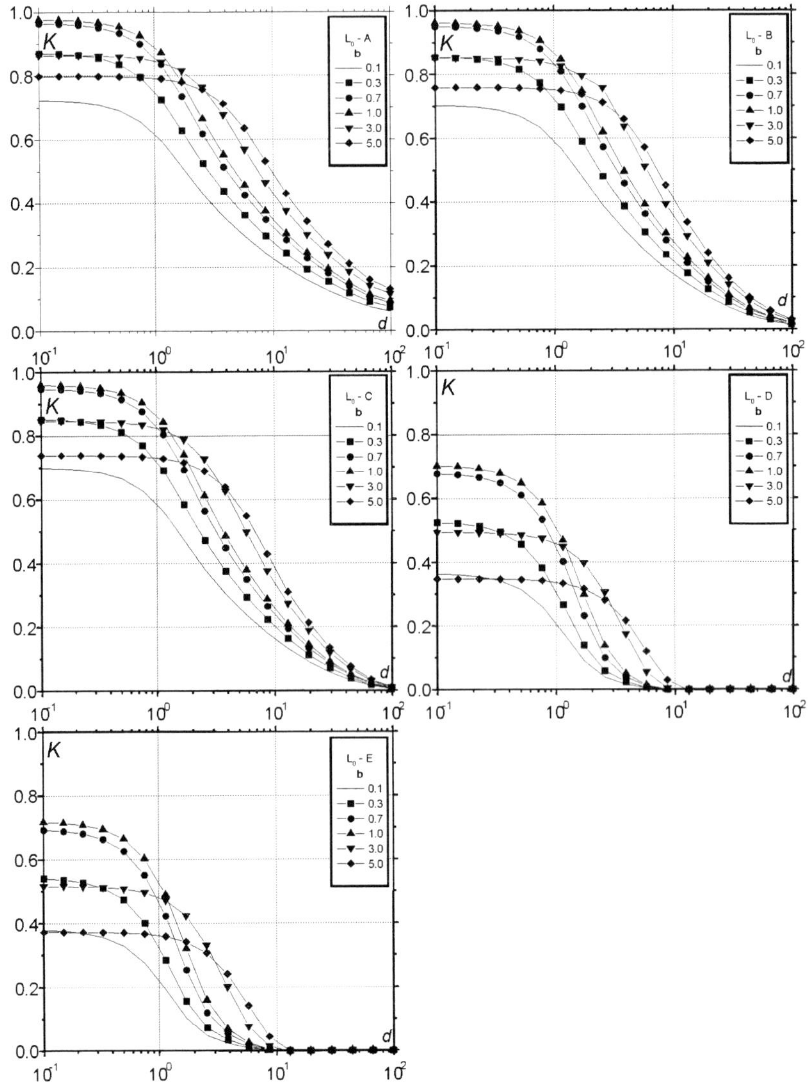

Figure 5.4.5. Correlation coefficient K from Eq. (5.4.7) for different models (A, B, C, D, E) of the outer scale of turbulence and the aperture radius of the laser system (parameter b); beacon altitude X=100 km.

5.4.2 Optimal algorithm for tip-tilt correction

The use of a laser guide star extends the domain of stable operation for an adaptive optics system. But the star is formed at some finite distance, so the problem arises of how to correct the data from optical measurements made against the guide star to ensure efficient compensation for aberrations in an image of a real astronomical object [23, 27]. This correction of the data becomes possible with the use of atmospheric models [52, 34, 36]. The models of atmospheric turbulence allow one to

- estimate the level of turbulent aberrations above the reference star, i.e., to calculate the optimal altitude of the star,
- compensate for partial focal isoplanatism in a system operating against a reference star situated at a finite altitude in the atmosphere, and
- improve the capabilities of tip-tilt correction.

Undoubtedly, when we use an LGS formed in the atmosphere by the backscattered signal in the algorithm correcting for random wandering [27, 28, 39, 34, 36] of the natural star image, the problem of optimization of data processing arises. Let us try to construct the algorithm for correcting the angular motion of a natural star image $\vec{\varphi}_F^{\mathrm{pl}}$ using the data measured on the LGS angular position $\vec{\varphi}_m$ as follows [34, 36]:

$$\vec{\varphi}_F^{\mathrm{pl}} - A\vec{\varphi}_m . \tag{5.4.10}$$

This algorithm should ensure minimization of the variance for residual angular displacements for the natural star image under tip-tilt correction with the proper choice of the coefficient A:

$$\langle\beta^2\rangle = \left\langle\left(\vec{\varphi}_F^{\mathrm{pl}} - A\vec{\varphi}_m\right)^2\right\rangle = \left\langle\left(\vec{\varphi}_F^{\mathrm{pl}}\right)^2\right\rangle + A^2\left\langle(\vec{\varphi}_m)^2\right\rangle - 2A\left\langle\vec{\varphi}_F^{\mathrm{pl}}\vec{\varphi}_m\right\rangle . \tag{5.4.11}$$

From Eq.(5.4.11) we obtain as a minimum:

$$\langle\beta^2\rangle_{\min} = \left\langle\left(\vec{\varphi}_F^{\mathrm{pl}}\right)^2\right\rangle - \left\langle\vec{\varphi}_F^{\mathrm{pl}}\vec{\varphi}_m\right\rangle^2 \Big/ \left\langle(\vec{\varphi}_m)^2\right\rangle , \tag{5.4.12}$$

when the correcting coefficient A is expressed through determinate functions as

$$A = \left\langle\vec{\varphi}_F^{\mathrm{pl}}\vec{\varphi}_m\right\rangle \Big/ \left\langle\varphi_m^2\right\rangle . \tag{5.4.13}$$

The form of this coefficient allows us to conclude that the coefficient A can be calculated from data in direct optical experiments (some time before adaptive telescope operation). Unfortunately, on the other hand, in a real experiment we can obtain only the data on $\vec{\varphi}_m$ because the vector $\vec{\varphi}_F^{\mathrm{pl}}$ characterizing angular wandering of the natural star image cannot be measured, owing to insufficient intensity of light from the natural star. In similar cases, we can estimate the coefficient A by Eq. (5.4.13), using the models of atmospheric turbulence [36] as well as the results of calculation of the correlation function [Eq. (5.4.4)] and the variances [Eqs. (5.4.5) and (5.4.6)].

It should also be noted that the minimum variance in Eq. (5.4.11) cannot be found traditionally by using the correction algorithm of Eq. (5.4.10) with $A = -1$. To demonstrate this, let us compare the residual variance for the optimal and nonoptimal (traditional) correction algorithms. In our designations, the minimum variance of residual fluctuations of angular shifts of the star image in the scheme shown in Fig. 5.3.1 can be estimated as

$$\left\langle \beta^2 \right\rangle_{\min} = \left\langle (\vec{\varphi}_F^{\,\mathrm{pl}})^2 \right\rangle \left(1 - \frac{2^{1/3} f(X,b,d,C_n^2)}{\left\{ 1 + b^{-1/3} - 2^{7/6}\left(1+b^2\right)^{-1/6} {}_1F_1\left[\frac{1}{6},\ 1;\ -\frac{d^2}{\left(1+b^2\right)}\right] \right\}} \right), \tag{5.4.14}$$

where the function

$$f(X,b,d,C_n^2) = \left(\int_0^X d\xi C_n^2(\xi)(1-\xi/X) \left\{ \begin{aligned} &[1+(1-\xi/X)^2]^{-1/6} \\ &-[1+b^2(1-\xi/X)^2]^{-1/6} \times {}_1F_1\left[\frac{1}{6}, 1; -\frac{d^2(1-\xi/X)^2}{1+b^2(1-\xi/X)^2}\right] \end{aligned} \right\} \right)^2 \times \left[\int_0^X d\xi C_n^2(\xi)(1-\xi/X)^{5/3} \int_0^\infty d\xi C_n^2(\xi) \right]^{-1} \tag{5.4.15}$$

depends on parameters of the optical experiment as well as on the atmospheric model used [Eqs. (5.4.10–5.4.15) are written assuming that the outer scale of turbulence is infinite and the LGS is a point source]. For the traditional algorithm [$A= -1$ in Eq. (5.4.10)], the residual variance for natural star motion is determined by Eq. (5.4.11).

Numerical analysis of these equations showed [34, 36] that the residual angular distortions in the optimal correction algorithm can be less than the residual distortions of traditional control methods. To illustrate the advantages of the optimal algorithm over the nonoptimal (traditional) one, we present Table 5.4.1, which gives the values of residual angular distortions for a telescope with a bistatic reference star. In this table, the values of residual variances are normalized to the variance of the angular motion of the natural star image without tip-tilt correction:

$$\left\langle \beta^2 \right\rangle \Big/ \left\langle \left(\vec{\varphi}_F^{\,\mathrm{pl}}\right)^2 \right\rangle = 1 + A^2 \left\langle (\vec{\varphi}_m)^2 \right\rangle \Big/ \left\langle \left(\vec{\varphi}_F^{\,\mathrm{pl}}\right)^2 \right\rangle - 2A \left\langle \vec{\varphi}_F^{\,\mathrm{pl}} \vec{\varphi}_m \right\rangle \Big/ \left\langle \left(\vec{\varphi}_F^{\,\mathrm{pl}}\right)^2 \right\rangle \tag{5.4.16}$$

Table 5.4.1. Comparison of the efficiency of optimal and traditional algorithms of tip-tilt correction for a bistatic scheme.

X (km)	b	Residual level of aberrations: Optimal algorithm	Residual level of aberrations: Traditional algorithm	A
8	0.3	0.640	1.291	-0.43
	0.5	0.603	1.105	-0.47
	0.7	0.578	0.999	-0.50
	1	0.552	0.899	-0.53
	2	0.500	0.736	-0.59
	3	0.471	0.656	-0.63
	5	0.434	0.570	-0.67
20	0.3	0.612	1.354	-0.42
	0.5	0.572	1.148	-0.46
	0.7	0.545	1.030	-0.49
	1	0.516	0.918	-0.52
	2	0.461	0.736	-0.58
	3	0.429	0.647	-0.62
	5	0.390	0.551	-0.66
40	0.3	0.602	1.406	-0.41
	0.5	0.561	1.187	-0.46
	0.7	0.533	1.062	-0.48
	1	0.504	0.944	-0.52
	2	0.447	0.751	-0.57
	3	0.414	0.657	-0.61
	5	0.374	0.556	-0.65
80	0.3	0.600	1.446	-0.41
	0.5	0.558	1.220	-0.45
	0.7	0.531	1.091	-0.48
	1	0.501	0.969	-0.51
	2	0.443	0.769	-0.57
	3	0.410	0.672	-0.60
	5	0.370	0.567	-0.64
100	0.3	0.599	1.455	-0.41
	0.5	0.588	1.227	-0.45
	0.7	0.530	1.097	-0.48
	1	0.500	0.974	-0.51
	2	0.443	0.774	-0.56
	3	0.410	0.676	-0.60
	5	0.370	0.570	-0.64

for the optimal and traditional algorithms of correction. The values of the correcting coefficient A calculated for the model of turbulence from Ref. [50] are given in the fifth column. The results are presented for different values of the size of the auxiliary laser beam (parameter b=0.3, 0.5, 0.7, 1, 2, 3, 5) and the LGS altitude (X= 8, 20, 40, 80, and 100 km); the model of the turbulent atmosphere was taken from Ref. [50] and the parameter d >5000.

It can be seen from the table that optimal correction with a properly chosen coefficient A decreases the residual distortions. So, a conclusion can be drawn about the efficiency of optimal correction using information on the altitude profiles of turbulence. At the same time, the intensity of distortions may even increase as a result of nonoptimal (traditional) correction in the bistatic scheme (for the parameter $d > 5000$, i.e., for a negligible level of correlation between the beam and motion of the natural star image), as can be seen from Figs. 5.4.1 and 5.4.2. But even with the considerable reduction of tilt jitter owing to the proper choice of A, a relatively large level of residual tilt affects the image of the natural star. It should be noted that this conclusion is based on the assumption that the LGS image is a point.

5.4.3 A laser guide star as an extended source

In some bistatic schemes, the LGS image has the form of an extended source [28, 38, 39]. As a consequence, random displacements for a motionless "secondary" source might be averaged over its angular extent [28, 38, 39]. This problem was considered early in Refs. [3], [5], and [29] for a reflecting surface and sources like a thin luminous string and an extended Gaussian beam.

In this section, the emphasis is on the difference and similarity between two bistatic schemes: Ragazzoni's approach [28] of two auxiliary slanted laser beams and a single telescope, and the approach used in papers [38] and [39], which is a main telescope with a laser beam and two auxiliary telescopes. In both schemes, the variances for random displacement of the LGS image are given by Eq. (5.1.1). If the observed LGS extent $a_b >> R_0$ (the aperture size of the main telescope or the aperture size of the auxiliary telescopes), then the variance [5, 29, 39] for the "secondary" source is described by the equation:

$$\left\langle \varphi_{ss}^2 \right\rangle = \left\langle (\varphi_F^{sp})^2 \right\rangle (a_b / R_0)^{-1/3} \frac{\int_0^X d\xi C_n^2(\xi)(1-\xi/X)^2(\xi/X)^{-1/3}}{\int_0^X d\xi C_n^2(\xi)(1-\xi/X)^{5/3}}. \quad (5.4.17)$$

In the next step, let us compare the minimal variances for the residual level of motion of the natural star image after tip-tilt correction for these two schemes with the optimal correction algorithm of Eq. (5.4.10). For the scheme from Ref. [28], the minimal variance is described by following equation:

$$\left\langle\left(\vec{\varphi}_F^{\rm pl} - A\vec{\varphi}_m\right)^2\right\rangle \Big/ \left\langle\left(\vec{\varphi}_F^{\rm pl}\right)^2\right\rangle = 1 - \left\langle\vec{\varphi}_F^{\rm pl}\vec{\varphi}_{\rm ss}\right\rangle^2 \Big/ \left[\left\langle\left(\vec{\varphi}_F^{\rm pl}\right)^2\right\rangle\left\langle(\vec{\varphi}_m)^2\right\rangle\right], \quad (5.4.18)$$

and for the scheme from Refs. [38] and [39]:

$$\left\langle\left(\vec{\varphi}_F^{\rm pl} - A\vec{\varphi}_m\right)^2\right\rangle \Big/ \left\langle\left(\vec{\varphi}_F^{\rm pl}\right)^2\right\rangle = 1 - \left\langle\vec{\varphi}_{\rm lb}\vec{\varphi}_F^{\rm pl}\right\rangle^2 \Big/ \left[\left\langle\left(\vec{\varphi}_F^{\rm pl}\right)^2\right\rangle\left\langle(\vec{\varphi}_m)^2\right\rangle\right]. \quad (5.4.19)$$

The last equation can be transformed into the form:

$$\left\langle\left(\vec{\varphi}_F^{\rm pl} - A\vec{\varphi}_m\right)^2\right\rangle \Big/ \left\langle\left(\vec{\varphi}_F^{\rm pl}\right)^2\right\rangle = 1 - K^2(0) \Big/ \left[1 + \frac{\left\langle\varphi_{\rm ss}^2\right\rangle}{\left\langle\varphi_{\rm lb}^2\right\rangle}\right], \quad (5.4.20)$$

where the correlation coefficient K is given in Eq. (5.4.7) for the parameter $d = 0$ and shown in Figs. 5.4.2–5.4.5. The second term in Eq. (5.4.20) is the normalized function of correlation between the angular image displacements of the plane wave and the secondary source—a broad beam having the size of the isoplanar zone in the LGS plane; as a result, it takes the form

$$\left\langle\left(\vec{\varphi}_F^{\rm pl} - A\vec{\varphi}_m\right)^2\right\rangle \Big/ \left\langle\left(\vec{\varphi}_F^{\rm pl}\right)^2\right\rangle = 1 - (2b)^{1/3} \frac{\left[\int_0^X d\xi C_n^2(\xi)(1-\xi/X)^{3/2}(\xi/X)^{-1/6}\right]^2}{\int_0^X d\xi C_n^2(\xi)(1-\xi/X)^{5/3}\int_0^\infty d\xi C_n^2(\xi)}. \quad (5.4.21)$$

The second terms in Eqs. (5.4.20) and (5.4.21) are similar, and hence it is possible to obtain an approximately equal level of correction in these approaches [27, 38, 39]. Using the models of the turbulent atmosphere [40, 41, 50] and applying Eqs. (5.4.20) and (5.4.21), we can estimate the minimal level of residual tip-tilt distortions for different LGS schemes.

The scheme of Refs. [38] and [39] requires two additional auxiliary telescopes to measure the motion of the LGS image with an accuracy of 0.05 arcsec. Hence, the approach in Ref. [28] is less expensive, owing to the use of only one large-scale telescope and two small-scale laser beam directors, whereas the second approach [38, 39] requires one main large-scale telescope with a laser beam setup and two auxiliary telescopes. The most promising method for full-aperture tip-tilt correction is to use an LGS adaptive optics system with a hybrid scheme and algorithms for tip-tilt correction [39, 57, 58].

5.5 Hybrid Scheme of Forming a Laser Guide Star

This section is logical continuation of a number of publications [34, 36, 59-61]. In addition, there is a definite connection between the main idea of this section and Refs. [62] and [39].

To implement the proposed correction algorithm based on the hybrid scheme of forming a laser reference star, three telescopes should be used: a main telescope and two auxiliary telescopes placed so that their configuration forms an isosceles rectangular triangle. The following scheme is used to form the laser reference star: A wide Gaussian laser beam is focused with the main telescope at a distance X. The star is formed solely by the central part of the main telescope (it is assumed that the initial laser beam diameter $a_0 < a_t$, where a_t is the aperture diameter of the main telescope).

In the focal plane of the main telescope, the angular jitter in the image centroid of the laser reference star is measured along the $0y$- and $0z$-axes. Simultaneously, in the focal planes of two auxiliary telescopes the angular shifts of the image along one of the two axes are measured in a direction transverse to the corresponding direction of separation of the axes of the main and auxiliary telescopes.

The laser reference star formed by focusing the laser radiation represents a long cylinder with a diameter a_m and length a_b ($a_b >> a_m$). Suppose that the separation between the main and auxiliary telescopes is such that for the auxiliary telescopes the laser reference star is formed by the bistatic scheme [60, 62, 39]. In this case, the size of a laser beacon a_b (connected with a_o, the altitude of star formation X, and separations between the axes of the auxiliary telescopes and the main telescope) seen from the location points of the auxiliary telescopes is much greater than the beacon size seen from the location of the main telescope ($a_b \gg a_m$). Thus, for the main telescope, the formed star can be considered to be monostatic. Then, the instantaneous position of its image (on the $0y$- and $0z$-axes) is

$$\begin{aligned} \varphi_{m,y} &= \varphi_{\text{lb},y} + \varphi_{\text{ps},y}\,, \\ \varphi_{m,z} &= \varphi_{\text{lb},z} + \varphi_{\text{ps},z}\,, \end{aligned} \qquad (5.5.1)$$

where ($\varphi_{\text{lb},y}$; $\varphi_{\text{lb},z}$) specify the instantaneous angular positions (on the axes) of the centroid of the laser beam focused at a distance X in the turbulent atmosphere; ($\varphi_{\text{ps},y}$; $\varphi_{\text{ps},z}$) specify the instantaneous angular positions of the image of the focused laser beacon considered as a point source. The auxiliary telescopes measure only one component of the image jitter of the laser reference star; that is, finally we have the following pair of measurable angles:

$$\begin{aligned} \varphi_{b,y} &= \varphi_{\text{lb},y} + \varphi_{\text{ss},y}\,, \\ \varphi_{b,z} &= \varphi_{\text{lb},z} + \varphi_{\text{ss},z}\,, \end{aligned} \qquad (5.5.2)$$

where ($\varphi_{ss,y}$; $\varphi_{ss,z}$) characterize the instantaneous angular positions of the image formed by an extended incoherent source (not fully correctly calculated in Ref. [39]). Furthermore, we calculate the corresponding differences:

$$\varphi_{m,y} - \varphi_{b,y} = \varphi_{ps,y} - \varphi_{ss,y}\,,$$
$$\varphi_{m,z} - \varphi_{b,z} = \varphi_{ps,z} - \varphi_{ss,z}\,. \tag{5.5.3}$$

Because the auxiliary telescopes operate in the regime of the bistatic reference star, the corresponding variances of differences [Eq. (5.5.3)] are expressed as

$$\left\langle(\varphi_{ps,y} - \varphi_{ss,y})^2\right\rangle = \left\langle(\varphi_{ps,y})^2\right\rangle + \left\langle(\varphi_{ss,y})^2\right\rangle = \left\langle(\varphi_{ps,y})^2\right\rangle\left\{1 + (a_b/a_{at})^{-1/3}\right\}, \tag{5.5.4}$$

where a_{at} is the size of the auxiliary telescope.

Now let us formulate the problem of optimal correction (decrease) for the angular jitter in the natural star $\vec{\varphi}_{ns}(\varphi_{ns,y}, \varphi_{ns,z})$ on the basis of the measured angles [Eqs. (5.5.1)–(5.5.3)] and the necessary calculations. In fact, we should minimize the variance of the residual angular shifts of the natural star through correction based on the measurements; namely,

$$\beta_y^2 = \left\langle[\varphi_{ns,y} - A(\varphi_{m,y} - \varphi_{b,y})]^2\right\rangle,$$
$$\beta_z^2 = \left\langle[\varphi_{ns,z} - A(\varphi_{m,z} - \varphi_{b,z})]^2\right\rangle. \tag{5.5.5}$$

Taking advantage of the results obtained in Refs. [60], [36], [61], and [34], we have (for the isotropic spectrum of turbulence)

$$\beta_y^2 = \beta_z^2 = \left\langle(\varphi_{ns,y})^2\right\rangle\left\{1 - \frac{2^{1/3} f(X, C_n^2)}{[1 + (a_b/a_{at})^{1/3}]}\right\}, \tag{5.5.6}$$

where

$$f(X, C_n^2) = \frac{\left\{\int_0^X d\xi C_n^2(\xi)(1-\xi/X)[1+(1-\xi/X)^2]^{-1/6}\right\}^2}{\int_0^X d\xi C_n^2(\xi)(1-\xi/X)^{5/3}\int_0^X d\xi C_n^2(\xi)}, \tag{5.5.7}$$

$\left\langle(\varphi_{ns,y})^2\right\rangle$ is the variance of the angular shift of the natural star image (along one axis); $b = a_0/a_{at}$. The optimal value of the correcting coefficient A minimizing the functionals [Eq. (5.5.5)] is calculated for the mean model vertical profiles of the structure parameter of the reflective index in the atmosphere $C_n^2(\xi)$ characterizing the turbulent intensity

$$A_{\mathrm{opt}} = \frac{2^{1/6} \int_0^X d\xi C_n^2(\xi)(1-\xi/X)[1+(1-\xi/X)^2]^{-1/6}}{\left[1+(a_b/a_{\mathrm{at}})^{-1/3}\right] \int_0^X d\xi C_n^2(\xi)(1-\xi/X)^{5/3}}. \tag{5.5.8}$$

Let us numerically estimate the efficiency of this correction for real experimental parameters. Let the main telescope have a diameter varying between 3 and 10 m. The auxiliary telescopes are from the class of 1-m telescopes. Let the diameter of the laser beam forming the star be $a_0 = 1$ m. The wave parameter for the focused laser beam Ω ($\Omega = ka_0^2/X$) ranges from 10 to 100 for altitudes X varying from 10 to 100 km. Hence, in the focal waist the size of the laser beacon is $a_{\mathrm{m}} = 1$–10 cm. Thus, the laser star cross section is seen by the main telescope at angles $\theta \leq 0.1$ arcsec and it can be practically considered as a point source. At the same time, the length of the laser star is a_b, and hence for proper separation of the auxiliary telescope axes, the visible size of the star a_b may be several minutes of arc; that is, the laser star can be considered as an extended incoherent source in the image planes of the auxiliary telescopes. In reality, the ratio $a_b/a_{\mathrm{at}} \approx 10^3$, $b = 1$. In calculations we used the mean model C_n^2 (ξ) suggested in Ref. [50].

Summarizing these data and making calculations, we obtain for Eqs. (5.5.6)–(5.5.8),

$$\Delta = \frac{\beta_y^2}{\langle(\varphi_{\mathrm{ns},y}^2)\rangle} = \frac{\beta_z^2}{\langle(\varphi_{\mathrm{ns},z}^2)\rangle} = \left[1 - \frac{2^{1/3} f(X, C_n^2)}{(1+0.1)}\right], \tag{5.5.9}$$

$$A_{\mathrm{opt}} = \frac{2^{1/6} \int_0^X d\xi C_n^2(\xi)(1-\xi/X)[1+(1-\xi/X)^2]^{-1/6}}{1.1 \times \int_0^X d\xi C_n^2(\xi)(1-\xi/X)^{5/3}}. \tag{5.5.10}$$

The results of numerical calculations, as well as the data for nonoptimal correction (that is, for $A = -1$) are given in Table 5.5.1. In the latter case,

$$\Delta = \frac{\beta_y^2}{\left\langle (\varphi_{\text{ns},y}^2) \right\rangle} = 1 + \left[1 + (a_b/a_{\text{at}})^{-1/3}\right] \frac{\int_0^X d\xi C_n^2(\xi)(1-\xi/X)^{5/3}}{\int_0^\infty d\xi C_n^2(\xi)}$$
$$-2^{7/6} \frac{\int_0^X d\xi \; C_n^2(\xi) \; (1-\xi/X) \; \left[1+(1-\xi/X)^2\right]^{-1/6}}{\int_0^\infty d\xi \; C_n^2(\xi)}. \quad (5.5.11)$$

Thus, it can be seen from the table that for a laser reference star higher than 10 km, this algorithm effectively corrects the jitter in the natural star image based on measuring the jitter (two components) in the monostatic star image in the main telescope and individual components of the jitter in two normally separated telescopes.

It should be noted that in practice there is no need to optimize the correction by this scheme (that is, to calculate the parameter A); nonoptimal correction (for $A = -1$) also highly efficiently corrects for the angular shift of the natural star image within the limits of isoplanar angles with the use of the laser reference star.

Table 5.5.1. Calculation results and data for nonoptimal correction.

X (km)	A_{opt}	Δ from Eq. (5.5.9)	Δ from Eq. (5.5.11)
1	−1.22	0.509	0.5139
10	−1.096	0.1799	0.1802
100	−1.019	0.0866	0.0927

5.6 Two Bistatic Schemes for LGS Formation

In this section, we present the results calculated for a "general" scheme of LGS formation in which it is possible to obtain an arbitrary value of correlation between random angular image displacements caused by fluctuations at the forward and backward paths.

As is known, the use of laser guide stars to improve the image quality in ground-based telescopes significantly extends the possibilities of applying adaptive correction in astronomy. However, the direct use of an LGS signal for full-aperture tip-tilt correction is impossible because it is first necessary to separate the valid signal from the data of optical observations on LGS image jitter.

In his recent papers Ragazzoni [28, 37, 62, 72-74] tried to systematize numerous approaches to solving the problem of LGS tilt retrieval for an adaptive optical system operating against an LGS signal. There are various approaches to solving of this problem; however, they also considerably increase the technical complexity of the system itself. For example, there are proposals to simultaneously measure the full-aperture angular jitter of a sufficiently bright natural reference star [73, 74] or to use two-color laser guide stars [75], additional telescopes [38, 39], or additional laser projectors [28, 37, 72]. In the last two cases, at a certain geometry the laser guide star cannot be considered as a point source. It should be noted that these approaches (with the use of additional telescopes or additional laser projectors) are realized through rather simple optical schemes.

Repeatedly in the scientific literature it was noted [28, 37, 39] that the bistatic schemes of LGS formation are the most efficient from the point of view of full-aperture tip-tilt correction. Figure 5.6.1 shows two proposed bistatic schemes. Scheme a has two additional telescopes (with the aperture R_a), and scheme b has two additional laser illuminators. For these schemes, the measured angular fluctuations of random displacement of the LGS image are [3]

$$\vec{\varphi}_m = \vec{\varphi}_{\text{lb}} + \vec{\varphi}_{\text{ss}}, \tag{5.6.1}$$

where $\vec{\varphi}_{\text{lb}}$ is the random vector of angular displacement of the laser beam forming the LGS and $\vec{\varphi}_{\text{ss}}$ is the random angular displacement of the image of the secondary source, actually an LGS.

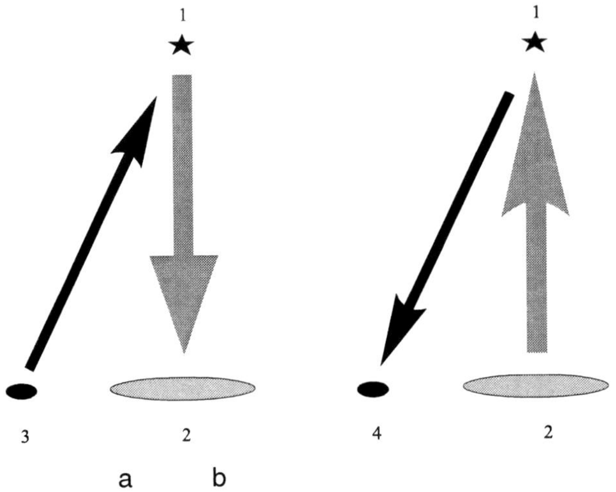

Figure 5.6.1. Bistatic schemes for the formation of a laser guide star: 1, laser guide star; 2, main telescope; 3, auxiliary laser; 4, illuminator auxiliary telescope. Scheme a was proposed in Refs. [28] and [37], scheme b in Ref. [39].

For the scheme a under conditions of a "limiting bistatic" [39, 8, 32, 34-36],

$$\left\langle (\vec{\varphi}_m)^2 \right\rangle = \left\langle (\vec{\varphi}_{\mathrm{lb}})^2 \right\rangle + \left\langle (\vec{\varphi}_{\mathrm{ss}})^2 \right\rangle. \tag{5.6.2}$$

Let us characterize the quality of tip-tilt correction as a ratio of the variance of residual fluctuations to the value of the initial variance $\left\langle (\vec{\varphi}_{\mathrm{pl}})^2 \right\rangle$:

$$\beta^2 = \left\langle (\vec{\varphi}_{\mathrm{pl}} - A\vec{\varphi}_m)^2 \right\rangle_{\mathrm{min}} / \left\langle (\vec{\varphi}_{\mathrm{pl}})^2 \right\rangle = \left\{ 1 - \frac{\left\langle \vec{\varphi}_{\mathrm{pl}} \vec{\varphi}_{\mathrm{lb}} \right\rangle^2}{\left\langle (\vec{\varphi}_{\mathrm{pl}})^2 \right\rangle \left\langle (\vec{\varphi}_m)^2 \right\rangle} \right\}. \tag{5.6.3}$$

If LGS is formed by the full aperture of the main telescope R_0, then $\left\langle \vec{\varphi}_{\mathrm{pl}} \vec{\varphi}_{\mathrm{lb}} \right\rangle = -\left\langle \vec{\varphi}_{\mathrm{pl}} \vec{\varphi}_{\mathrm{sp}} \right\rangle$, where $\vec{\varphi}_{\mathrm{sp}}$ is the angle of displacement of the point source image in the focal plane of the main telescope. We obtain from Eq. (5.6.3) [36, 58, 51, 66, 67]:

$$\beta^2 = 1 - \frac{K^2(X)}{1 + \left\langle (\vec{\varphi}_{\mathrm{ss}})^2 \right\rangle / \left\langle (\vec{\varphi}_{\mathrm{sp}})^2 \right\rangle}, \tag{5.6.4}$$

where

$$K(X) = \frac{\left\langle \vec{\varphi}_{\mathrm{pl}} \vec{\varphi}_{\mathrm{sp}} \right\rangle}{\sqrt{\left\langle (\vec{\varphi}_{\mathrm{pl}})^2 \right\rangle} \sqrt{\left\langle (\vec{\varphi}_{\mathrm{sp}})^2 \right\rangle}}.$$

For the models in Refs. [40], [44], [68], [41], [42], [69], and [70] for turbulence parameters, this last function is given in Table 5.6.1 for various heights X of formation of the laser guide star.

Table 5.6.1. Function K for various heights of LGS formation.

X (km)	$K(X)$	$K^2(X)$	1- $K^2(X)$
1	0.65	0.42	0.58
2	0.72	0.52	0.48
3	0.75	0.56	0.44
4	0.79	0.62	0.38
10	0.84	0.71	0.29
20	0.88	0.77	0.23
40	0.89	0.79	0.21
100	0.90	0.81	0.19

At the same time, for scheme b, the result of tip-tilt correction is

$$\beta^2 = 1 - \frac{\left\langle \vec{\varphi}_{\text{pl}} \vec{\varphi}_{\text{ss}} \right\rangle^2}{\left\langle (\vec{\varphi}_{\text{pl}})^2 \right\rangle \left[\left\langle (\vec{\varphi}_{\text{lb}})^2 \right\rangle + \left\langle (\vec{\varphi}_{\text{ss}})^2 \right\rangle \right]} . \tag{5.6.5}$$

It can be easily shown that for a real height X the correlation $\left\langle \vec{\varphi}_{\text{pl}} \vec{\varphi}_{\text{ss}} \right\rangle = \left\langle \vec{\varphi}_{\text{pl}} \vec{\varphi}_{\text{sp}} \right\rangle$. Thus, the efficiency of full-aperture tip-tilt correction for the bistatic scheme [51] is determined by the value of $K^2(X)$ and the ratio $\left\langle (\vec{\varphi}_{\text{ss}})^2 \right\rangle / \left\langle \vec{\varphi}_{\text{sp}}^2 \right\rangle$. For further consideration, it is necessary to take into account the size of the secondary source and calculate the jitter of an extended source.

5.6.1 Jitter of an extended source

It is known that in bistatic schemes the LGS is not a point object. If the longitudinal size of the LGS is designated as l_b, then the observed size of a beacon at points separated from the axis of the main telescope by (ρ^x, ρ^y) is

$$a_b^z = l_b \rho^z / X , \quad a_b^y = l_b \rho^y / X .$$

Let us consider numerically the variance of the jitter of the extended source image by comparing it with the jitter image of a spherical wave. Let us designate their attitude

$$\left\langle (\vec{\varphi}_{\text{ss}})^2 \right\rangle / \left\langle (\vec{\varphi}_{\text{sp}})^2 \right\rangle = f(b,a). \tag{5.6.6}$$

The variance of the jitter of an extended source image was studied rather thoroughly earlier by a number of authors [3, 5, 29, 37]. We use the equation for $\left\langle (\vec{\varphi}_{\text{ss}})^2 \right\rangle$ from Ref. [39] with small corrections. Then, to within 5% accuracy we obtain for the ratio

$$\begin{aligned} \left\langle \varphi_{\text{ss}}^2 \right\rangle / \left\langle \varphi_{\text{sp}}^2 \right\rangle = & \int_0^X d\xi C_n^2(\xi)(1-\xi/X)^2 \Big\{ \left[a_b^2 (\xi/X)^2 + R_a^2 (1-\xi/X)^2 \right]^{-1/6} \\ & - \left[a_b^2 (\xi/X)^2 + R_a^2 (1-\xi/X)^2 + 2\kappa_0^{-2} \right]^{-1/6} \Big\} \\ & \times \left[\int_0^X d\xi C_n^2(\xi)(1-\xi/X)^2 \Big\{ \left[a^2 (1-\xi/X)^2 \right]^{-1/6} - \left[a^2 (1-\xi/X)^2 + 2\kappa_0^{-2} \right]^{-1/6} \Big\} \right]^{-1} . \end{aligned} \tag{5.6.7}$$

Here, a_b is seen as the linear size of an extended "secondary" source, and R_a is the size of the aperture of an additional telescope. We use here the following normalizations: $b = a_b / R_0$, $a = R_a / R_0$, and $c^2 = 2\kappa_0^{-2} R_0^{-2}$.

The ratios of variances $\left\langle (\vec{\varphi}_F^{ss})^2 \right\rangle / \left\langle \vec{\varphi}_{lb}^2 \right\rangle = f(b,a)$ for various ratios of a_b, R_a, a_0 were calculated for a model of a spectrum of atmospheric turbulence [40, 44, 68, 41, 42]:

$$\Phi_n(\kappa,\xi) = 0.033 C_n^2(\xi) \kappa^{-11/3} \left[1 - \exp(-\kappa^2/\kappa_0^2)\right],$$

where $\kappa_0^{-1}(\xi)$ is the outer-scale atmospheric turbulence. In the calculations in Ref. [51] the outer scale of atmospheric turbulence was set as a constant size for the whole atmosphere. This is hardly justified for all cases. Let us take advantage of several models [69, 70] of a vertical structure of the outer scale (taking into account models A–E from Sec. 5.3).

The model of high-altitude profiles of $C_n^2(\xi)$ was taken from Ref. [31]; appropriate "average" conditions of vision through an atmosphere are used. Some models for describing the high-altitude profiles of outer-scale atmospheric turbulence presented here are used; namely, models C and E, along with fixed values of κ_0^{-1} = 3, 10, 100, and 1000 m. The calculations were carried out for several sizes of apertures of the main telescope: R_0 = 1, 4, and 10 m. The results of the calculations of functions of $\left\langle (\vec{\varphi}_F^{ss})^2 \right\rangle / \left\langle \vec{\varphi}_{lb}^2 \right\rangle = f(b,a)$ are presented in Figs. 5.6.2. and 5.6.3.

The results are given in Fig. 5.6.2 (for $X = 10$ km) and Fig. 5.6.3 (for $X = 100$ km) as six graphs. The graphs on the left vertically correspond to R_a / R_0 = 0.1 and those on the right to R_a / R_0 = 1.0. Horizontally, the three groups correspond to main telescope's aperture sizes: R_0 = 1 m, 4 m, and 10 m. The ratio of the observed size of an extended "secondary" source a_b to the initial size of a laser beam was set in intervals from 0.1 to 1000. Thus, we have all the characteristics necessary to estimate the efficiency of correction of a general wavefront inclination.

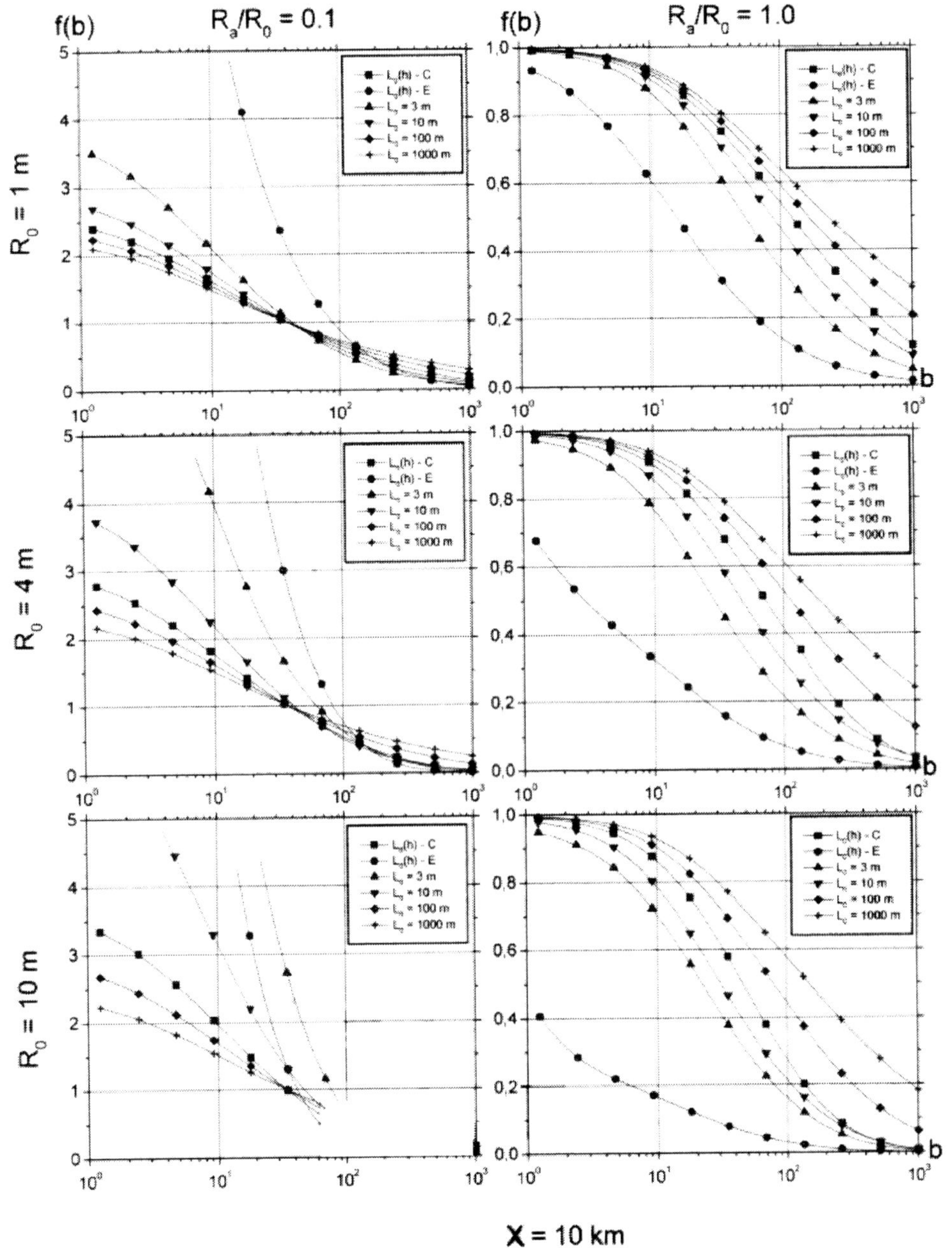

Figure 5.6.2. Variance of the image jitter of an extended source obtained by comparing it with the image jitter of a spherical wave. The altitude of the reference laser star was 10 km.

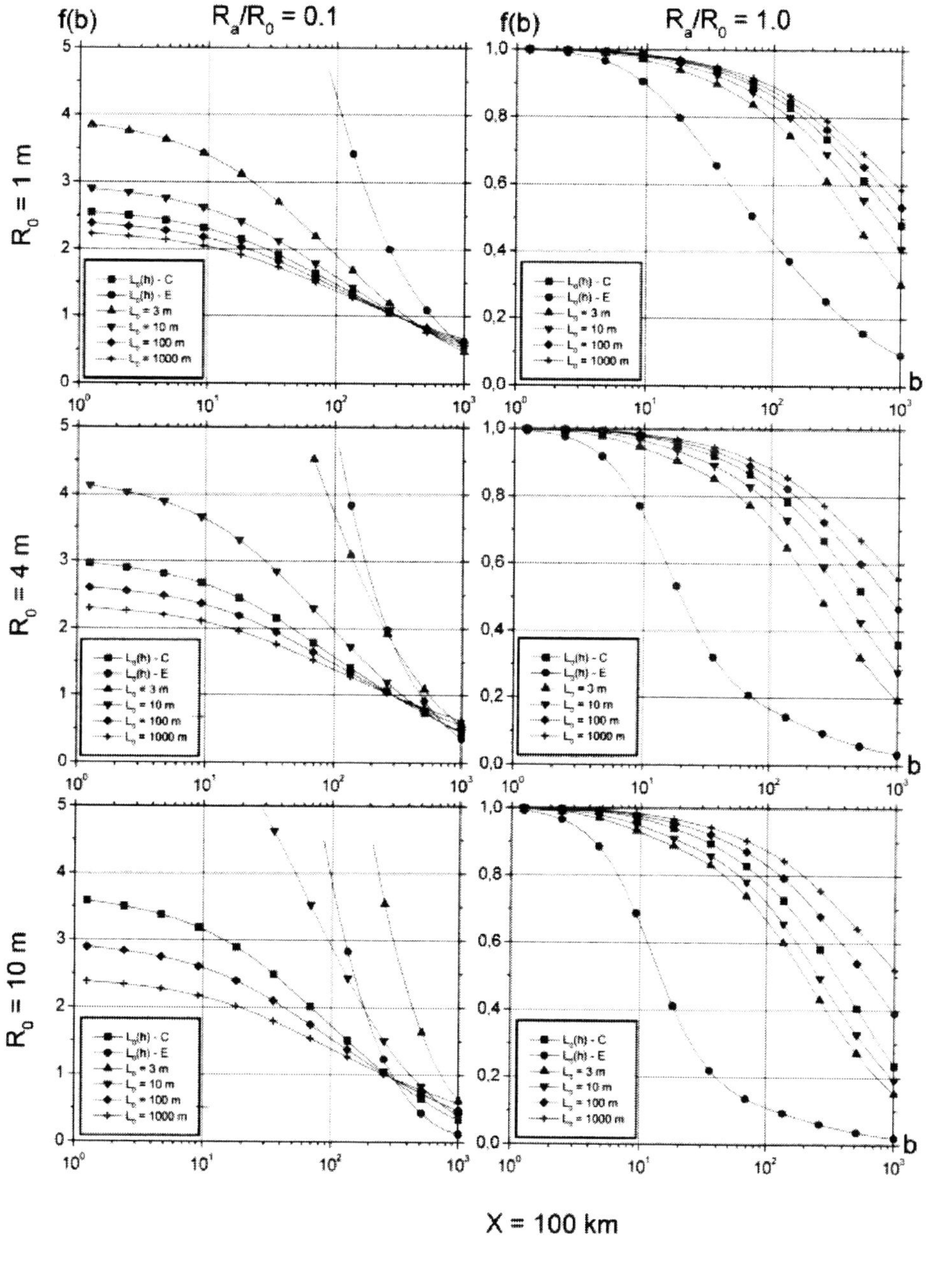

Figure 5.6.3. Variance of the image jitter of an extended source obtained by comparing it with the image jitter of a spherical wave. The altitude of the reference laser star was 100 km.

5.6.2 Basic shortcomings of schemes a and b for LGS formation

Using the designations of Eq. (5.6.6) we have for scheme a:

$$\beta^2 = 1 - \frac{K^2(X)}{1 + f(b,a)}, \tag{5.6.8}$$

and for scheme b:

$$\beta^2 = 1 - \frac{K^2(X)}{f(0,\ a) + f(b,\ 1)}. \tag{5.6.9}$$

Note here that $f(0,a)$ characterizes the altitude of the variances for a point source for the apertures R_a and R_0, and for $f(0,\ 1) = 1$. Thus, using our calculations (Table 5.6.1, and Figs. 5.6.2 and 5.6.3), proceeding from the formulas of Eqs. (5.6.8) and (5.6.9), it is possible to obtain the relative efficiency of correction of a general wavefront inclination. The following requirements should be noted:

For scheme a:

1. At formation LGS by the full aperture of the main telescope has a place the phosphorescence of elements of an optical train of a telescope.
2. Besides the main telescope, it is necessary to have two rather large auxiliary telescopes.
3. Two additional wavefront sensors are also needed.
4. The spacing of the axis of an auxiliary telescope should be at least 10–40 km.

For scheme b:

1. It is necessary to have a laser illuminator of a rather large size.
2. The spacing of the axes of the telescope and illuminator should be on the order of 10–40 km.

It is possible to ensure a high level of correction of the fluctuations of a general wavefront inclination in observatories with smaller telescopes if the largest telescope is used to create the guide star.

5.6.3 A differential scheme

Here we present a new approach: the differential scheme for LGS. This scheme uses (see Fig. 5.6.4) a narrow laser beam formed outside of the aperture (and even of the dome) of the main telescope and two additional telescopes. Jitter of the image is measured in the aperture of the main telescope:

$$\varphi^{y}_{\text{mono}} = \varphi^{y}_{\text{lb}} + \varphi^{y}_{\text{ss}}(R_0), \quad \varphi^{z}_{\text{mono}} = \varphi^{z}_{\text{lb}} + \varphi^{z}_{\text{ss}}(R_0)$$

and the auxiliary telescopes

$$\varphi^{y}_{\text{bi}} = \varphi^{y}_{\text{lb}} + \varphi^{y}_{\text{ss}}(R_a), \quad \varphi^{z}_{\text{bi}} = \varphi^{z}_{\text{lb}} + \varphi^{z}_{\text{ss}}(R_a).$$

We calculate their differences:

$$\varphi^{y}_{\text{mono}} - \varphi^{y}_{\text{bi}} = \varphi^{y}_{\text{ss}}(R_0) - \varphi^{y}_{\text{ss}}(R_a), \quad \varphi^{y}_{\text{mono}} - \varphi^{y}_{\text{bi}} = \varphi^{y}_{\text{ss}}(R_0) - \varphi^{y}_{\text{ss}}(R_a). \tag{5.6.10}$$

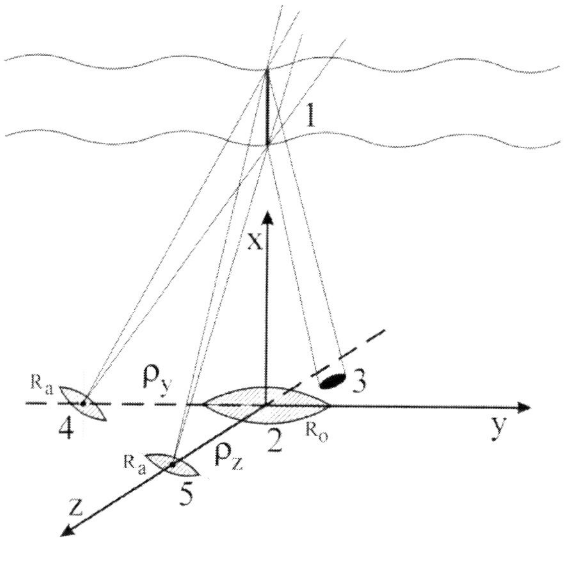

Figure 5.6.4. A differential scheme for LGS formation with an auxiliary illuminator and two auxiliary telescopes.

Since the main telescope "sees" a guide star as a point, i.e., $\vec{\varphi}_{\text{ss}}(R_0) = \vec{\varphi}_{\text{sp}}(R_0)$, the efficiency of a full-aperture tip-tilt correction for this differential scheme is

$$\beta^2 = 1 - \frac{K^2(X)}{1 + f(b,a)}. \tag{5.6.11}$$

The advantages of such a differential scheme are (1) there is no dependence on parameters of the laser transmitter, and (2) there is no parasitic illumination of elements of the main telescope aperture.

To take advantage of temporary averaging [71] of the differential signal of Eq. (5.6.10), we obtain

$$\beta^2 = 1 - \frac{K^2(X)}{1 + f(b,a)g(T)}, \quad (5.6.12)$$

where the function $g(T) < 1$ and T is the period of temporary averaging of the differential signal.

At the same time, as can be seen from Figs. 5.6.2 and 5.6.3, a good correction level for a general wavefront inclination requires a scheme in which $b = 10^3$, i.e., $a_b = 10^3 R_0$. This means that the spacing of the axes should be more than

$$\rho^y > 10^3 R_0 X / l_b, \; \rho^z > 10^3 R_0 X / l_b. \quad (5.6.13)$$

For R_0 = 4 m, X= 100 km, l_b =10 km, we obtain 40 km. It is probably not convenient to have too large a space between the main and auxiliary telescopes.

5.7 A New Scheme for LGS Formation

Here we present a new, dynamic type of LGS. It is possible to implement a different technique to create an LGS as two crossing lines. In this scheme, production of two narrow laser beams and fast angular scanning of each of these beams (in two orthogonal directions) are carried out simultaneously. The rapid scanning creates a luminous cross.

In a monostatic receiver, a wavefront sensor (in the focal plane of the main telescope) based on two CCD matrices is used to detect signals from two lines: one along the y and a second along the z direction. The optical train of the wavefront sensor allows us to obtain only one image of the single line on each CCD matrix. Special controls for each matrix are used to obtain two signals from each matrix; namely, signal φ^y from entire y matrix (or φ^z from the z matrix) and a signal φ_c^y from the middle part of the y matrix (or φ_c^z from the z matrix). We used the differences of these signals:

$$\Delta^y = \varphi^y - \varphi_c^y \text{ for the } y \text{ matrix and } \Delta^z = \varphi^z - \varphi_c^z \text{ for the } z \text{ matrix.}$$

It is easy to show that the components of this equation are

$$\varphi^y = \varphi_{\text{lb}}^y + \varphi_{\text{ss}}^y, \quad \varphi_c^y = \varphi_{\text{lb}}^y + \varphi_{\text{sp}}^y,$$

$$\varphi^z = \varphi_{\text{lb}}^z + \varphi_{\text{ss}}^z, \quad \varphi_c^z = \varphi_{\text{lb}}^z + \varphi_{\text{sp}}^z.$$

Finally, the differential signal as a vector is $\vec{\Delta} = \vec{\varphi}_{ss} - \vec{\varphi}_{sp}$. The residual level of the general inclination for a common correction is

$$\beta^2 = \frac{\langle [\vec{\varphi}_{pl} - \vec{\Delta}]^2 \rangle}{\langle (\vec{\varphi}_{pl})^2 \rangle} \approx \frac{\langle [\vec{\varphi}_{pl} - \vec{\varphi}_{sp}]^2 \rangle}{\langle (\vec{\varphi}_{pl})^2 \rangle} + f(b,1).$$

The first term is connecting with the cone nonisoplanarity, and second term with extension of the LGS. Note that for an optimal algorithm of correction, the residual level of general inclination is

$$\beta^2 = \frac{\langle [\vec{\varphi}_{pl} - \vec{\Delta}]^2 \rangle}{\langle (\vec{\varphi}_{pl})^2 \rangle} = 1 - \frac{K^2(X)}{1 + f(b,\ 1)}.$$

Through the choice of the angular scanning value, the second components of the differential signal can be made as small as needed. This value should be enough to ensure high-frequency (10–20 kHz) scanning of an angle on the order of 2–3 degrees of an arc and equivalent to spacing of an auxiliary telescope axis at 40 km. As a result, additional telescopes are not required; it is only necessary to have a wavefront sensor with two matrices in the structure of the main telescope. Thus, it is necessary to ensure that the field of view of this wavefront sensor is approximately 2–3 degrees of an arc (with a telescope with an 8-m diameter, it is necessary to have a focus of no more than 200 m).

5.7.1 Attempt to reduce the effect of angular anisoplanatism

Between an observable star and another star (chosen as guide star) let there be an angular mismatch equal to θ. It is known that any function (let it be phase fluctuations observed in the zenith star) can be represented as the decomposition of orthogonal polynomials,

$$S(y,\ z,\ 0) = \sum_{1}^{N} a_j F_j(y/R,\ z/R).$$

Let us take advantage of the fact that the correlation functions $b_j(\theta)$ of the fluctuations of the modal components $a_j (j = 1, ..., N)$ can be calculated. Further data on measurements carried out on the guide star can be corrected (in the θ direction) in the following scheme:

$$S(y,\ z,\ \theta) = \sum_{1}^{N} b_j(\theta) a_j F_j(y/R,\ z/R)\,. \tag{5.7.1}$$

It is easy to show that for any one modal component, the residual mistake caused by focal nonisoplanarity is expressed in a component of the decomposition of a phase as follows:

$$\left\langle \left[a_j(0) - a_j(\theta)\right]^2 \right\rangle = D_j(\theta) = 2\left\langle a_j^2 \right\rangle [1 - b_j(\theta)]\,. \tag{5.7.2}$$

It is obvious that at angles of θ exceeding the angular of correlation, the error is reduced twofold. If we take advantage of an estimate like that of Eq. (5.7.1), we find that the residual mistake for any component of Eq. (5.7.2) becomes only

$$\left\langle \left[a_j(\theta) - a_j(\theta)\right]^2 \right\rangle = \left\langle a_j^2 \right\rangle [1 - b_j^2(\theta)]\,.$$

With an increased number of modal components, the radius of correlation also falls, and we find that at the fixed angle θ of a mismatch we can more than double the accuracy of the correction.

References

1. J.C. Dainty, C.J. Solomon, N.J. Wooder, A.N. Bogaturov, A.S. Gurvich, V.A. Myakinin, "Double passage imaging," in meeting digest of "*Scintillation*" International Meeting for Wave Propagation in Random Media, editors V.I. Tatarskii, A.Ishimaru, 1992.
2. V.A. Banakh, V.L. Mironov, *Propagation of Laser Sensing Radiation in the Atmosphere,* Nauka, Novosibirsk, 1986.
3. V.M. Orlov, I.V. Samokhvalov, G.G. Matvienko, M.L. Belov, A.N. Kozhemyakov, *The Elements of Theories of Wave Scattering and Optical Ranging,* Nauka, Novosibirsk, 1982.
4. Meeting digest of "*Scintillation*" International Meeting for Wave Propagation in Random Media, editors V.I. Tatarskii, A. Ishimaru, 1992.
5. M.A. Kalistratova, A.I. Kon, "Fluctuations of the arrival angle of light waves from an extended source in a turbulent atmosphere," *Izv. V.U.Z. Radiofiz.* **9**, No. 6, pp.1100–1107, 1966.
6. I. Smith, T. Pries, K.I. Skipka, M.A. Hinuiter, "High-frequency plane-wave filter function for folded paths," *J. Opt. Soc. Am.*, **62**, No. 9, pp.1183–1187, 1972.
7. I. Smith, T. Pries, "Temporal-frequency spectra for waves propagating over straight and folded paths," *Appl. Opt.*, **14**, No. 5, 1975.

8. V.P. Lukin, "Tracking of random angular displacements of optical beams," in *Proceedings V All-Union Symposium*, Part II, pp. 33–36, 1979.
9. V.P. Lukin, "Correction for random angular displacement of optical beams," *Sov. J. Quantum Electron.*, **10**, No.6, pp.727-732, 1980.
10. V.P. Lukin, "The fluctuation of wave beams in a turbulent atmosphere under reflection," *Proceedings I All-Union Meeting on Atmospheric Optics*, Part I, pp.129–133, 1976.
11. V.P. Lukin, V.V. Pokasov, "The experimental measurements of fluctuations in backscattering wave in a turbulent atmosphere," in *Proceedings I All-Union Meeting on Atmospheric Optics*, Part I, pp.134–137, 1976.
12. V.P. Lukin, "Efficiency of the compensation of phase distortions of optical waves," *Sov. J. Quantum. Electron.*, **7**, No. 4, pp. 522–524, 1977.
13. V.P. Lukin, "Phase fluctuations of specularly reflected optical waves," *Opt. i Spektrosk.* **48,** No. 3, pp. 594–599, 1980.
14. V.P. Lukin, V.V. Pokasov, S.M. Slobodjan, "The correction of random angular displacements of wave beams," in *Proceedings I All-Union Meeting on Atmospheric Optics*, Part I, pp.259–262, 1976.
15. V.P. Lukin, V.M. Sazanovich, S.M. Slobodjan, "Random displacements of images at sensing in the atmosphere," *Proceedings V All-Union Symposium on Atmospheric Laser Ranging*, Part II, pp. 43–48, 1978.
16. V.P. Lukin, "Registration of random angular displacements of optical beams," in *Proceedings V All-Union Symposium on Laser Beam Propagation*, Part II, pp. 33–36, 1979.
17. V.P. Lukin, "Improvement of an image quality formed through the turbulent atmosphere," in *Proceedings II All-Union Meeting on Atmospheric Optics*, Part II, pp. 93–96, 1980.
18. O.N. Emaleev, V.V. Pokasov, S.F. Potanin, V.P. Lukin, V.M. Sazanovich, O.L. Tuzov, "Fluctuation of images in the close-to-earth layer of the atmosphere," in *Proceedings II All-Union Meeting on Atmospheric Optics*, Part II, pp. 97–100, 1980.
19. V.I. Klyatskin, *Statistical Description of Dynamical Systems with Fluctuating Parameters*, Nauka, Moscow, 1975.
20. V.I. Tatarskii, *Wave Propagation in a Turbulent Atmosphere*, Nauka, Moscow, 1967.
21. V.P. Lukin, V.F. Matyuchin, "An adaptive image correction," *Sov. J. Quantum Electron.*, **13**, No. 12, pp. 1604–1610, 1983.
22. V.P. Lukin, M.I. Charnotskii, "The reciprocity principle and adaptive control of optical radiation parameters," *Sov. J. Quantum Electron.*, **12,** No. 5 , pp. 602–606, 1982.
23. V.P. Lukin, *Atmosfernaia Adaptivnaia Optika,* Nauka, Novosibirsk, 1986.
24. V.P. Lukin, *Atmospheric Adaptive Optics,* SPIE Press Vol. PM 23, 1996.
25. D.L. Fried, "Least-squares fitting a wavefront dislocation estimate to an array of phase-difference measurements," *J. Opt. Soc. Am.*, **67**, No. 3, pp. 370–375, 1977.

26. V.P. Linnik, "On the possibility of reducing the influence of atmospheric seeing on the image quality of stars," *Opt. Spectrosc.*, **47,** No. 4, pp. 401–402, 1957.
27. R. Fugate, "Laser beacon adaptive optics," *Opt. Photonics News*, **4,** pp. 14–19, 1993.
28. R. Ragazzoni, "Absolute tip-tilt determination with laser beacons," *Astron. Astrophys.* **305**, L13-L16, 1996.
29. V.L. Mironov, V.V. Nosov, B.N. Chen, "Correlation of shifting of laser source optical images in a turbulent atmosphere," *Izv.V.U.Z. Radiofiz.*, **25**, No. 12, pp. 1467–1471, 1982.
30. V.P. Lukin, "Efficiency of compensation of phase distortion of optical waves," *Sov. J. Quantum Electron.*, **4**, pp. 923–927, 1977.
31. V.P. Lukin, O.N. Emaleev, "Correction of angular displacements of optical beams," *Sov. J. Quantum Electron.*, **9**, pp. 2264-2271, 1982.
32. V.P. Lukin, B.V. Fortes, "Efficiency of adaptive correction of images in a telescope using an artificial guide star," *OSA Tech. Digest*, **23**, pp. 92–93, 1995.
33. V.P. Lukin, "Limiting resolution of an adaptive telescope with the use of an artificial star," *Proc. OSA*, ICO-16, *Active and Adaptive Optics*, pp. 521–524, 1993.
34. V.P. Lukin, "Laser beacon and full aperture tilt measurement," *OSA Tech. Digest*, **13**, Addendum AMB-35, pp. 1–5, 1996.
35. V.P. Lukin, "Adaptive formation of beams and images in a turbulent atmosphere," *Atm. Oceanic Opt.*, **8**, pp. 301-341, 1995.
36. V.P. Lukin, B.V. Fortes, "Comparison of limit efficiencies for various schemes of laser reference star formation," *Atm. Oceanic Opt.*, **10**, pp. 34–41, 1997.
37. R. Ragazzoni, S. Esposito, E. Marchetti, "Auxiliary telescopes for absolute tip-tilt determination of a laser guide star," *Mon. Not. R. Astron. Soc.*, **276**, pp. L76–L78, 1995.
38. M.S. Belen'kii, "Full aperture tilt measurement technique with a laser guide star," *Proc. SPIE Vol.* **2471**, pp. 289–296, 1995.
39. M.S. Belen'kii, "Tilt angular correlation and tilt sensing techniques with a laser guide star," *Proc. SPIE Vol.* **2956**, pp. 206–217, 1996.
40. R.E. Good, R.R. Beland, E.A. Marphy, J.H. Brown, E.M. Dewan, "Atmospheric models of optical turbulence," *Proc. SPIE Vol.* **928,** pp.165–186, 1988.
41. V.P. Lukin, "Investigation of some peculiarities in the structure of large-scale atmospheric turbulence," *Atm. Oceanic Opt.*, **5,** pp. 834–840, 1992.
42. V.P. Lukin, "Intercomparison of models of the atmospheric turbulence spectrum," *Atm. Oceanic Opt.*, **6,** pp.1102–1107, 1993.
43. V.V. Voitsekhovich, "Outer scale of turbulence: Comparison of different models," *J. Opt. Soc. Am. A.*, **12**, pp. 1346–1354, 1995.

44. T.S. McKechnie, "Atmospheric turbulence and the resolution limit of large ground-based telescopes," *J. Opt. Soc. Am. A.*, **9**, pp. 1937–1954, 1992.
45. M. Sarazin, ed., *Site Testing for the VLT (Very Large Telescope)*, European Southern Observatory, VLT Report No. 50, 1990.
46. N. Nakato, M. Iye, I. Yamaguchi, "Atmospheric turbulence of small outer scale," European Southern Observatory Conference and Workshop Proceeding, F. Merkle, ed., No. 48, pp. 521–524, 1993.
47. A. Agabi, J. Borgino, F. Martin, A.V. Tokovinin, and A. Ziad, "G.M.S: A grating scale monitor for atmospheric turbulence measurements. II. First measurements of the wavefront outer scale at the O.C.A," *Astron. Astrophys. Suppl. Ser.*, **109**, pp. 557–562, 1995.
48. J. Borgino, "Estimation of the spatial coherence outer scale relevant to long baseline interferometry and imaging in optical astronomy," *Appl. Opt.* **29**, No. 13, pp. 1863–1865, 1990.
49. V.P. Lukin, E.V. Nosov, B.V. Fortes, "Effective outer scale of atmospheric turbulence," *Atm. Oceanic Opt.*, **10**, pp. 100–106, 1997.
50. M.A. Gracheva, A.S. Gurvich, "Simple models of turbulence," *Izv. Akad. Nauk SSSR, Fiz. Atmos. Okeana*, **16**, pp.1107–1111, 1980.
51. V.P. Lukin, "Monostatic and bistatic schemes and an optimal algorithm for tilt correction in a ground-based adaptive telescope," *Appl. Opt.*, **37**, pp. 4634–4644, 1998.
52. V.P. Lukin, "Models and measurements of atmospheric turbulence characteristics and their impact on AO design," *OSA Tech. Digest*, **13**, pp.150–152, 1996.
53. R.J. Sasiela, J.H. Shelton, "Transverse spectral filtering and Melin transform technique applied to the effect of outer scale on tilt and tilt anisoplanatism," *J. Opt. Soc. Am. A.* **10,** pp.646–660, 1993.
54. C.E. Coulman, J. Vernin, Y. Coquegniot, and J.L. Caccia, "Outer scale of turbulence appropriate to modeling refractive-index structure profiles," *Appl. Opt.*, **27,** pp. 155–160, 1988.
55. *Site Testing for the VLT. Data Analysis Part I,* European Southern Observatory, VLT Report No. 55, 1987.
56. *Site Testing for the VLT. Data Analysis Part II*, European Southern Observatory, VLT Report No. 60, 1990.
57. *Yearly Status Report on Adaptive Optics at the Telescopio Nationale Galileo,* edited by R. Ragazoni, 1996.
58. V.P. Lukin, "Hybrid scheme of formation of a laser reference star," *Atm. Oceanic Opt.*, **10,** pp. 975–979, 1997.
59. V.P. Lukin, B.V. Fortes, "Efficiency of adaptive correction images in telescope using an artificial guide star," in *ESO Conference and Workshop Proceeding*, No. 54, pp. 191–194, 1995.
60. V.P. Lukin, "Laser guide stars and problem of full aperture tilt measurements," *Atm. Oceanic Opt.*, **9**, No. 11, pp. 948–956, 1996.

61. R. Ragazzoni, "Propagation delay of a laser beacon as a tool to retrieve absolute tilt measurements," *Astrophys. J.*, **465**, pp. L73–L75, 1996.
62. R.Q. Fugate, W.J. Wild, "Untwinkling the stars—Part I.," *Sky Telescope*, pp. 25–32, 1994.
63. V.P. Lukin, B.V. Fortes, *Adaptive Formation of Beams and Images in the Atmosphere*, SB RAS Press, Novosibirsk, 1999.
64. V.P. Lukin, "Distinction and similarity of two schemes of formation of a laser guide star," *Atm. Oceanic Opt.*, **11**, N.11. pp. 1253–1257, 1998.
65. V.P. Lukin, "Optical measurements of the outer scale of atmospheric turbulence," *Atm. Oceanic Opt.*, **5**, pp. 354–377, 1992.
66. V.P. Lukin, "Some problems of use of laser guide stars," *Proc. SPIE Vol.* 3983, pp. 90–100, 1999.
67. S. Esposito and R. Ragazzoni, "Technique for LGS tilt retrieval: A numerical comparison," in *Adaptive Optics at the Telescopio Nazionale Galileo*, pp. 1–20, 1997.
68. S. Esposito, R. Ragazzoni, "Nonconventional technique for LGS tilt retrieval: An update," in *Adaptive Optics at the Telescopio Nazionale Galileo*, pp. 1–20, 1997.
69. R. Ragazzoni, "Robust tilt determination from a laser guide star using a combination of different techniques," *Astron. Astrophys.*, No. 319, pp. L9–L12, 1997.
70. R. Foy, A. Migus, F. Biraben, G. Gryberg, et al., "The polycromatic artificial sodium star: A new concept for correcting atmospheric tilt," *Astron. Astrophys.*, No. 111, pp. 599–578, 1995.

CONCLUSION

The range of problems that are successfully solved with the use of laser systems has grown considerably in recent years. The expanding application of current optoelectronic and adaptive optics systems is state-of-the-art optics. Historically, adaptive optics was first used in astronomy and more recently in ground-based systems for imaging artificial satellites and other space objects, but now other efficient applications are being reported increasingly often. In our opinion, in the near future adaptive optics will be widely used in a number of areas. Current adaptive optics can produce breakthroughs in industrial technologies and medicine. The performance of optoelectronic devices in manufacturing and medical applications (such as welding and cutting, drilling metal and extrahard materials, laser scalpels, or optical systems in ophthalmology) that employ coherent processing of signals can be considerably improved through implementation of adaptive optics elements and systems.

However, the development of optoelectronic systems is a rather long process that usually proceeds in the absence of complete information on the peculiarities of the medium of propagation. Also, modern optoelectronic systems are rather expensive because at the initial stage of their design they require calculating and estimating the efficiency of applying various algorithms and programs that are based on current adaptive optics technologies. The main parameters of such systems can hardly be changed during their operation. This forces designers to keep in mind possible changes in the system's main parameters at different stages of design and production.

The feasibility of sufficiently flexible adjustment of optical system parameters is now one of the main requirements in designing optoelectronic devices, and this feasibility is provided by adaptive optics systems. This is because a change in the main parameters of an optical system can be made simply by replacing the operational algorithm of an adaptive system.

The creation of optical systems necessarily includes designing the systems and determining the possibilities of their application under actual atmospheric conditions. In this book we have described these stages, taking into account present-day achievements.

Adaptive optics systems differ from other systems in the elements included in the optical scheme: a wavefront sensor, an active optical element (active mirror), and a reference source supplying information on fluctuations and the radiation propagation channel. Each element of the optical scheme calls for precalculations; therefore, calculations of not only the parameters of adaptive optics, but the system as a whole are often needed. Corresponding calculations are performed for the wavefront sensor, active mirror, and certainly the reference source formed in the propagation channel of the optical radiation to be corrected.

One of the ways to create a reference source is to use a signal backscattered from atmospheric inhomogeneities. Various schemes for formation of a reference source (a laser guide star) for imaging purposes have been described. In this book, we discussed the limited capabilities of image correction using a signal based on measurement of the LGS position. The direct use of an LGS signal for full-aperture tip-tilt correction is impossible because the valid signal should be first separated from the data of optical observations of LGS image jitter. This aspect of the problem was described in detail in this book with allowance made for modern advances in this field. Earlier, attempts were undertaken to systematize numerous approaches solving the problem of LGS tilt retrieval for an adaptive optics scheme operating against an LGS signal. We have considered here a "general" scheme for forming a laser guide star.

Since numerical simulation of optical wave propagation through the atmosphere is now one of the main methods for studying and designing modern optoelectronic systems, we have paid considerable attention to describing the computational algorithms that can be used in software packages for modeling the adaptive control of laser beams and imaging systems in the atmosphere. The propagation of optical radiation through a randomly inhomogeneous medium is simulated using a numerical solution of the wave equation written in the parabolic approximation for the scalar complex amplitude of the optical field and the field of the refractive index of the medium. When dynamic and nonlinear problems are modeled, the wave equation is solved together with some material equation describing how the state of the medium changes in time. In our calculations we use the modified splitting method and the fast Fourier transform algorithm.

The algorithms and programs developed by us allow the operation of an optoelectronic system to be modeled as a whole. It becomes possible to describe such phenomena as nonstationary thermal blooming, in which the refractive index of the medium varies because the medium is heated by the laser radiation propagating through it. The algorithms proposed here make it possible to model the evolution of the temperature field by taking into account two mechanisms: forced convection (in the arbitrary direction of the wind velocity) and molecular heat conductivity; this is very important if there are dead zones on the path of the laser radiation.

To take into account the effect of turbulent fluctuations of the atmospheric refractive index on the propagation of laser radiation (the laser beam), we model two-dimensional randomly inhomogeneous phase distortions of the wavefront with a spectral density corresponding to the Kolmogorov model of the turbulence spectrum that accounts for the finiteness of the inner and outer scales of turbulence. According to the Kolmogorov–Obukhov hypothesis, the structure function in fluctuations of temperature and the atmospheric refractive index obeys the power law. The finite values of the inner and outer scales of turbulence were introduced in the calculations. The ratio of these scales is taken, as a rule, to be equal to 1000. The structure function depends on the intensity of turbulent distortions. In the atmospheric surface layer ($h < 20$ m) this intensity decreases with height, and the character of this height dependence changes with meteorological conditions. This significantly complicates the development of a single universal model. We use several models, including a rather simple empirical model obtained from experimental data (up to 20 km) under the conditions of best, medium, and worst visibility.

The conditions of propagation of laser radiation through the atmosphere include such characteristics as the position of the laser source and the position and motion of the receiver. The atmosphere is modeled as a stratified medium. The variable parameters of the problem are the altitude of the laser source, the initial radius of the beam, the radiation intensity at the optical axis, the intensity distribution profile of the optical beam, the altitude of the receiver, the zenith angle of the propagation path, and the azimuth and scanning rate of the laser source.

To take into account the vertical variability of atmospheric parameters entering into the equations to be solved, we propose to use the standard models of the atmosphere that allow for physical and geographical conditions, as well as spatiotemporal variations of meteorological parameters based on statistical measurements over many years.

The atmospheric air is assumed to be an ideal gas of a constant composition that is described by the state equation, including pressure, density, and temperature. The atmosphere is divided into the following layers: troposphere, stratosphere, mesosphere, and thermosphere. The altitude profile of the temperature for each layer is approximated by a linear function of the geopotential altitude. The vertical profile of the air density is calculated from the given profiles of temperature and pressure based on the state equation of the ideal gas.

Because of considerable spatiotemporal variability of the wind in the atmosphere, when solving applied problems we think it is worth using the data from online sensing of the path along which the optoelectronic system will operate. However, to evaluate the efficiency of adaptive optics systems designed to operate through the atmosphere, it is quite sufficient to restrict consideration to the models of the wind structure that were obtained from long-term measurements at sensing stations.

Since molecular absorption of laser radiation in the atmosphere has pronounced frequency dependence, a line-by-line calculation is now the most universal and accurate method for determining absorption characteristics.

Actual adaptive optics systems employ several types of sensors, including the Hartmann sensor, to record phase distortions. We propose a specialized sensor based on the Hartmann algorithm. We use up-to-date approaches to develop a stable phase reconstruction algorithm, in particular for strong intensity scintillations, as well as algorithms for phase "joining" under dislocation conditions.

We have analyzed a wide range of control elements with different degrees of freedom, geometries of mutual arrangement, and frequencies and spatial fluctuations of the response of every control element. In this analysis we included mirrors for tip-tilt correction of the optical beam as a whole. In addition, we have studied various active and adaptive mirrors: zonal and modal correctors and segmented mirrors of different geometries, as well as a static model of a flexible mirror and a numerical model of a dynamic mirror.

The authors hope that this book will be interesting for its readers. It will be useful for specialists dealing with the development of devices and elements of adaptive optics systems.

INDEX

About the Authors

VLADIMIR P. LUKIN is Director of the Wave Propagation Division and the Head of the Laboratory of Coherent and Adaptive Optics at the Institute of Atmospheric Optics, Siberian Branch, Russian Academy of Sciences. He is also a part-time professor at Tomsk State University's Department of Radiophysics. His principal field of work is application analysis, stochastic fields, adaptive optical system theory, and electromagnetic wave propagation. Professor Lukin has received the 1987 RAS Award for outstanding fundamental contributions to the theoretical sciences of nonlinear atmospheric optics, and the 1988 RAS Award for outstanding applied contributions for his work on "Calculations of the influences on the operation of laser opto-electronics systems." Since 1994 he has received several Russian State grants that are awarded to outstanding researchers in atmospheric optics, two scientific grants from the International Science Foundation, and a grant from the European Southern Observatory. Prof. Lukin received a "For Merit for Country" II medal degree in 1999, and the International Commission for Optics awarded him the 2000 Galileo Galilei Award. He has served as Vice-editor of the international scientific journal *Atmospheric and Oceanic Optics*; in 1998, he was a guest coeditor of a special issue "Adaptive Optics" for the journal *Applied Optics.* In addition, he has published 8 monographs (including *Atmospheric Adaptive Optics,* SPIE Press, 1996), 7 patents, and 216 research publications in refereed journals, for a total of more than 380 scientific papers in all.

BORIS V. FORTES, a graduate of Tomsk State University (1988, Physics), is employed as a Senior Researcher for the Laboratory of Coherent and Adaptive Optics at the Institute of Atmospheric Optics, Siberian Branch, Russian Academy of Sciences. He is also a part-time professor in the Department of Radiophysics at Tomsk State University. His principal field of work is numerical simulation for adaptive optical system theory and problems of electromagnetic waves propagation. Fortes has authored more than 80 scientific papers, including one monograph.